Executive's Guide
to the Wireless Workforce

by
Russell D. Lambert

Edited by
Julia King
Computerworld

WILEY

John Wiley & Sons, Inc.

384.33
L22e

Library of Congress Cataloging-in-Publication Data

Lambert, Russ D.
  Executive's guide to the wireless workforce / Russ D. Lambert
      p. cm.
Includes index.
  ISBN 0-471-44879-6
  1. Cellular telephone systems. 2. Wireless communication systems. I. Title: Guide to the
wireless workforce. II. Title.
  HE9713.L36 2003
  384.3'3'068--dc21   2003000574

Printed in the United States of America

10 9 8 7 6 5 4 3 2 1

Contents

Preface

Wireless will change many things, probably more than we can imagine, and it will happen one practical application at a time.

Wireless. It was less than 100 years ago that we began building a telecommunications infrastructure and a new economy that relied on wires. Today, most executives are convinced that wireless computing will have a significant impact on the two core economic fundamentals—customer interaction and business processes. But those same executives are hesitant to believe the incessant technology hype. We cringe at the claims of reinventing your business with the new technology *du jour*, especially when many of the people who made those claims are out of business.

Without question wireless computing has begun to significantly penetrate the business mainstream. By 2004 there will be more than a billion devices in the world that can remotely access the Internet or a corporate network. In the United States, more than half of all businesses will make their critical software applications available wirelessly to their employees or customers by 2006, when more than half the workforce will be mobile in some way.

Tens of thousands of executives will be charged with developing and implementing a wireless strategy for their company. To make that happen, they'll spend an estimated $4.7 billion by 2005. You need to know about wireless computing now. What this book does is provide nontechnical executives (e.g., CEO, CFO, VP Sales) with case study examples of how over 50 companies across multiple industries deployed wireless to employees and customers. This book is a compendium of their lessons learned, best practices, and potential business benefits gained from wireless computing as well as a step-by-step methodology for understanding, analyzing, and developing a wireless strategy that fits your company.

This book is not a version of wireless for dummies, because that approach is mostly a how-to method for beginning technophiles. It is more accurately wireless for the rest of us in business. If you are a CEO or senior executive, this book will help you understand the core issues of business-related wireless computing with minimal technical layers. If you are an IT executive, you may find a nugget or two that helps you, but you would benefit more from giving this book to your CEO to aid in understanding the issues and potential of wireless for the enterprise—then propose your budget for wireless!

This book can assist every business executive in any size company to gain a better understanding of the technical challenges, strategy elements, and the commercial issues incumbent in using wireless as a tool for business.

The "Wireless Workforce," "Wireless Warehousing and Logistics," "Wireless in Health Care, Government, and Education," and "Wireless for Customers" chapters are packed with examples of enterprises using wireless in practical proven ways. There are more than 50 cross-industry case studies of companies like Fidelity Investments and McKesson, as well as smaller companies like Bartlett Tree Experts, all of which have stepped out and applied a wireless technology to specific business processes creating a sustainable competitive advantage. These First Practice 50 enterprises can serve as a catalyst to spark ideas in your company. What if you identify a few ideas that have promise—what then?

The "Mapping Your Company's Wireless Strategy" and "Extending the Enterprise to Wireless" chapters provide a strategic framework and easy-to-navigate templates and checklists for planning and executing an enterprisewide wireless strategy. Use these chapters as guides to incorporate wireless technology into your company's internal and customer-facing business processes. Use the "Top Ten Wireless Lessons Learned" to avoid the biggest pitfalls in wireless deployments.

The last two chapters, "Wireless Forces of Change" and "Looking Ahead," provide an overview of the macro factors and practical innovations in motion that will affect the next 10 years of our business future. These are not intended as a futuristic forecast, but more of an acknowledgment of anticipated near-term products, services, and trends that will impact the bottom line.

Speaking of the bottom line—you need to know about wireless! Let's get started!

RUSS LAMBERT

Acknowledgment

Writing a book is an exercise in perseverance. Liken it to running a mental marathon. As such, you need encouragement at every mile and I am thankful to have received it from my family: Sarah, Beth, James, Joshua, and Mary. Early guidance from my trusted friend Randall Dobbins and an insightful edit from Mark Hepburn were invaluable. Most importantly, this work could not have been accomplished without the initial and continuing energetic and thoughtful contributions by Julia King, whose command of the business technology issues is surpassed only by her editing acumen. Thanks Julia. We did it.

Introduction

The single most important thing you need to know about wireless is that it will affect your company's earnings. Exactly how and precisely when may be unknown, but it is certain. It's only a matter of time before wireless is used in some way by most businesses, fundamentally changing the way that employees, customers, and suppliers interact. Quietly, without press releases or Super Bowl ads, companies everywhere are rethinking and optimizing their business processes using wireless communications and mobile computing applications. The evidence is compounding. FedEx, UPS, and Frito Lay were early pioneers of wireless computing. Then came Fidelity, Thrifty Car Rental, and Sears, which have demonstrated successful application of wireless and mobile computing across multiple business models. Today the list of companies applying wireless continues to grow— with more than 50 case studies included in this book alone.

Wireless will not reinvent your business or allow you to dominate your competitors with a single stroke of innovation. Regardless of technology, the fundamentals of good business practice haven't changed that much; however, the pace and pressure of today's business environment is bordering on oppressive. To stay ahead of the competition and to meet customers' ever-rising demands for immediate, anytime-anywhere service, instant information, and of course low prices, you must consider every material variable that can affect your company's success. We all may be a little numbed by the blizzard of technology claims and constant feature mania, but take this as a wakeup call. You need to know about wireless.

You will learn in this book how dozens of companies, both large and small, are using wireless to cut costs, improve productivity, and grow their businesses. You will learn how to approach the strategic issues involved in deploying wireless computing and how to manage through the tactical challenges.

The fax machine increased the velocity of paper interactions to enhance business communication and transactions at a speed and with precision never experienced before (electronic paper). Computers allowed the rapid aggregation, calculation, and presentation of key business data that has forever changed the way we manage, operate, and communicate in business (electronic processing). Wireless computing has the profound potential to dramatically transform, or at the least significantly enhance, certain business processes and customer interactions, leading us to the next stage of productivity evolution—the electronic worker. You can't afford to miss the opportunity to lower your costs of operation, deepen your interactions with your customers, or get a lot smarter about your business faster. The company that recognizes these possibilities, does its homework, and then applies the technology economically and practically will clearly benefit. Lee Iacocca put it this way: "The most successful businessman is the one who holds onto the old just as long as it is good, and grabs the new just as soon as it is better." The facts speak strongly to the point that wireless computing can now be considered "better" than ever before, but when is the right time to commit to the technology?

Working in e-commerce for the last few years, I described our strategy as "the second mouse gets the cheese." What this meant was allowing other companies to lead out with a huge investment in the hottest software or process, then after it was proven and a few more vendors entered the fray, lowering the price levels, we would jump in. We missed the opportunity for a few press releases, but we did not overspend. There is wisdom in waiting, at least a little while, for an emerging technology to really emerge. The 50 case study references in this book are clear evidence that the waiting time is over for business wireless computing—it's time to jump in!

What's driving companies to begin investing in wireless is first and foremost, efficiency. The word is getting out. Real use of wireless computing is increasing in the business space, driven by the promise of increased efficiency and productivity. Transportation, health care, distribution, insurance, construction, and service companies are all quietly adopting wireless technology to automate or transform their business processes and their customer interactions. Think about it. Most likely, more than a few companies you're acquainted with—as a consumer, business partner, or supplier—are deploying wireless or mobile computing as you read this book.

Think back on your reading over the last several months. Chances are good that you saw or read more than a few articles on business and wireless. If you began saving the business and trade journal articles about companies that are announcing the use of wireless devices to improve some business process, you might be surprised as you watch the file grow. You would quickly confirm the accelerating trend toward deployment of wireless computing technology as a competitive strategy for many companies.

The poster child for deploying wireless is FedEx, whose digital-assisted dispatch system is able to efficiently route thousands of drivers for more efficient pickup and delivery. By coupling the dispatch system with the shipper's SuperTracker handheld device that communicates location and shipping information to a central system, FedEx effectively transformed itself into a real-time organization. Both innovations created substantial economies, including a 20% increase in the number of packages delivered each day by the thousands of FedEx drivers. On the back end, their customer service group eliminated the need to write down a million addresses a day. FedEx was far ahead of the pack in 1995, investing in a big way and taking a big risk on brand-new technology. Thanks primarily to their wireless deployments, FedEx and UPS have both boosted the efficiency of their truck fleets and drivers by 20%, accomplishing a sustainable leap in performance. The difference today is that the risks and cost of entry for wireless are lower, and the outcomes are more certain.

Skepticism may linger about consumers' adoption of wireless, but the idea of applying wireless to traditional business processes is steadily building. IDC, a technology research firm, polled large and small companies about their plans for wireless computing in summer 2001. Even then the survey revealed that one in five (21%) of the respondents already had some basic form of wireless computing capability in place, and 43% of the companies were considering deploying a solution. The early adopters in the study were weighted to larger companies with heavy representation in health care, government, and utilities, but subsequent studies show wireless computing initiatives moving to almost every industry segment.

This book profiles more than 50 companies and organizations across every industry that has deployed wireless computing. I call them the First Practice 50, and they help prove that wireless is moving to the business mainstream—and it will move even faster.

FROM NOVELTY TO NORM

The personal computer (PC) is arguably the granddaddy of the technology revolution, but it is only a youthful twenty-something. Even younger is the Internet, a technology that very few of us were using as recently as the early 1990s. Widespread use of e-mail, banking over the Internet, buying airline tickets online, and personal digital assistants (PDAs) only became widely available around 1992, but both the PC and the Internet are now considered essential to business communications. Can you imagine not having e-mail to get your job done? (Except for the junk mail!) We often take for granted the relative newness of transforming technologies that have been rapidly and broadly integrated into our collective business processes. What was a novelty in the recent past is relied on today as a critical and indispensable business tool. That's where wireless computing is headed.

The long-term effect of the "e"-everything cycle experienced in the late 1990s ultimately will reflect more than a single technology, business idea, or speculative business model. The early speculative failures of e-business models or volatile consumer phenomena are by no means the endgame; rather, they are part of a continuing wave of technological evolutions that are improving business processes and customer service. Placing the computer, Internet, and wireless technologies into a timeline reveals that we are only at the beginning of a larger and longer-term wave of business process optimization that taps the powerful communications and distributed computing technologies of all three.

Using computers with wireless capabilities to extend the reach of computing and the Internet (hereafter called "wireless") is the next logical step in a truly profound march of technological advancements. Everything we know about the benefits of computing, together with explosive use of the Internet to improve the flow of information and communications between and among people and businesses, applies to the potential of wireless. It is not a stretch to believe that wireless computing is quickly moving from novelty to norm.

WISDOM OF WAITING

Despite the growing list of companies deploying wireless, many companies have taken a wait-and-see attitude toward the technology. There is a healthy respect for the perceived past constraints in wireless technology such as speed, limitations in devices, and uncertain return on investment (ROI) of technology in general. Broad acceptance of new technology by a generally conservative business community requires demonstrable evidence that the new technology pass a basic test: New technology must add more value than it costs to acquire, deploy, and operate. This is a fairly high hurdle to leap, but it makes a lot of sense. Waiting until this test is met usually pays off. The rare exception is when competitive pressure together with customer pressure prompts earlier investment. Barnes & Noble did not have a choice when Amazon entered the online book business. Ready or not, Barnes & Noble had to respond or concede the online market. A similar competitive squeeze may occur with wireless in your industry, but the basic test must be met, especially in light of the hype and misguided investment in the late 1990s.

In today's ROI-based economy, we all come from Missouri—a "show me" sign hangs over every board meeting reviewing technology investment requests. Yet despite the disappointments in e-business and Internet ventures, practical wireless innovation persists.

According to industry research in late 2001, 66% of American companies plan to make their critical software applications available wirelessly within the next three years to either their employees or their customers or both. On the consumer side, by 2004, the value of wireless consumer mobile commerce, dubbed m-commerce, is projected to be $21 billion, with some 177 million consumers using the wireless Web to purchase everything from airline tickets and stocks to apparel and furniture. To take advantage of this huge market opportunity, Datamonitor is projecting that U.S. corporations will be spending an average $4 to 5 billion per year on wireless computing solutions by 2005.

Growing evidence suggests that wireless computing is meeting the ROI test and fueling employee productivity gains. Evidence also shows an emerging competitive imperative to deploy wireless applications in customer relationship and customer service applications. Overall, a compelling demand is building for mainstream use of wireless computing by large and small businesses. The decision you face now is whether to defer using wireless until you read about a competitor's innovation in the news or to dig in and get to work with wireless now.

GETTING STARTED

Incorporating wireless technology into a company's internal and customer-facing business processes is not overly complex, but it is also not simple or straightforward. We are early on in development and adoption of wireless for business, and there are no clear methodologies or industry practices to benchmark. As a result, companies are developing wireless applications by trial and error.

This book will help by providing nontechnical managers with a step-by-step methodology for understanding, analyzing, and developing a wireless strategy that fits their company. The guide includes succinct and straight-forward steps for defining, assessing, and deploying wireless technology. Throughout the book are more than 50 cross-industry case study refer-ences that specifically show how companies such as Fidelity Investments, McKesson Corp., and Sears have deployed wireless to gain a competitive advantage. Learning from their wireless experiences may spark some ideas for your business.

Doing your homework on this subject takes effort. This is not one of those books that is futuristic and fuzzy—you know, the kind you read and think you need to do something, but you just don't know what you need to do! This book includes practical how-to chapters that are full of check-lists, examples, and lessons learned.

I have attempted to avoid excessive use of buzzwords, acronyms, and consultant-ese, but unfortunately, like a lot of technologies, wireless has a specific set of definitions and labels you should review. You need to know enough of the terminology to understand the solutions and benchmark your competitors. To help, I've provided the necessary glossary of terms and acronyms written in plain English.

REFLECTING ON THE CHAPTER

All executives should know and appreciate that at some point in the next decade, every company will be faced with competitive pressures to deploy wireless in some manner. The impact to your bottom line may occur sooner than you think. When your market intelligence teams inform you that your competitor is offering your customers wireless services and information that you can't offer, the cost and time pressure to play catchup will be sub-stantial. That may be occurring as you read this book.

A great CEO I worked for would often say: "The most difficult thing we do is to think!" This appropriately applies to the subject of wireless in this fast-paced and sometimes confusing technology-based business environ-ment. In my experience, even very complex business scenarios can be understood and leveraged with sufficient study, discussion, and thoughtful action. I hope this book helps you with that thinking process.

When you finish this book, you won't necessarily know more about technology than the consultants or the wireless specialists in your IT group, but you will have a holistic view with some clear strategies and tactics to deploy. A careful review of the First Practice case examples will provide you with benchmarks to both educate and bolster your confidence in finding a workable ROI for wireless in your organization.

Most important, you will be able to carry on an informed conversation and think through the important issues with your technology support teams. Take your place at the planning board. It is time to plan how your company will utilize the next and possibly the most important wave of business innovation—wireless.

1

Wireless Basics for Business

This chapter provides a background orientation for those who need to review the fundamentals about wireless. If you are new to the subject of wireless, this chapter should be useful because it includes a basic definition of wireless and wireless computing, a discussion of the evolution of wireless networks and mobile devices, and a brief summary of extending the corporate operating system (the enterprise system) to wireless devices. This chapter also reviews the different roles and players involved in the business of wireless, from the big companies such as Microsoft and the network providers, to the hundreds of midsized players that manufacture phones or provide software. There are also more than enough consultants to help us sort it all out. When you finish this chapter, you should have a good overview of the various elements in the wireless game. The subjects briefly touched on here are examined in greater detail throughout the book.

WHAT IS WIRELESS ANYWAY?

Wireless is a catchphrase for communications (both voice and data) without the use of wires. Signals are transmitted over invisible radio waves instead of copper or fiber-optic cable. Garage door openers and television remote controls were among the first wireless devices. Next came the breakthrough of wireless phones for voice transmission. Today we have a variety of devices equipped to use the mobile phone network to accomplish wireless computing. Wireless computing is the extension of information networks—including the Internet and corporate intranets—to remote devices such as personal digital assistants (PDAs), pocket PCs, and tablet PCs. Wireless computing allows for the near full-function use of a computer without being hardwired to a network. So far, practical use of the wireless communications networks for true mobile computing has been slow, but it is steadily growing.

Transmitting voice and small portions of data over the existing telecommunications networks works pretty well; however, accomplishing wireless computing over the telephone network has been much more challenging. In fact, until recently, it hasn't worked well at all. That's why almost every major telecommunications company has an infrastructure upgrade project underway (in various stages of completion and coverage) across the entire country. Their goal is to augment wireless voice communications with high-speed data transmission, which supports high speed wireless computing. This task is obviously important to the future of wireless computing and has the potential, together with some other critical factors, to have a profound effect on everyone.

Like the PC speed wars in the 1990s, the network carriers are competing to offer the next generation of faster data networks for the anticipated millions of wireless data users waiting for greater speed. Every carrier is on a quest to deliver a nationwide, always-on, high-speed wireless data service. Each wireless carrier is hoping to offer faster Web browsing and expanded data-handling capacity to its customers better and cheaper than the other guys.

To keep track of the rapid changes and varying capabilities of the networks, there are several labels for various communication methods (protocols) and the associated throughput rates. Broad definitions of classes of network characteristics (like speed) are assigned the label "G," which is short for "generation." First-generation (1G) communications capabilities are through hardwired phone lines moving analog (not digital) information over traditional copper wires. Innovations like cable modems and digital subscriber lines (DSL) within the 1G environment allow data to move even faster over the existing copper networks.

The second generation of telecommunications and data transmission is done without wires (hence, wireless) and is called 2G. This generation of network evolution includes several primary types: GSM, CDMA, TDMA, and an isolated standard called iDEN. Think of the different methods of achieving 2G as different engines driving the same results. There is no agreement on what standard is best or most cost effective, but everyone agrees that 2G is not fast enough to move data at a rate that will make Web browsing, file retrieval, and viewing video practical for hand-held wireless devices. It must be faster. It must be 3G. Well, maybe.

The third genertaion offers a bigger "pipe" for more information to travel faster. It increases throughput speeds between 10 to 100 times faster than 2G. Third-generation network deployments began in 2001, and by 2002 were in commercial use in many large cities. The promise of what 3G networks will enable is mind-boggling: 3G will allow you to use a wireless device with the same connectivity and capabilities as your desktop computer. Imagine watching a football game on your PDA, or more practically having a real-time interactive video teleconference over your hand-held. You can surf the Web in real time, viewing images and downloading large files with no latency (slowness).

Now for the confusing part: There are various levels of throughput speeds between the 2G and 3G standards; two of them are called 2.5G and 3Glite. They achieve substantial increases in network performance at lower cost and are being deployed faster than 3G. Achieving nationwide consistent coverage with compatible standards and coverage is the catch that comes with the innovation. Planning the development and deployment of your company's wireless application must include careful consideration of the evolution in network throughput speeds and devices that work with those networks in your operating geography. Why? Because you don't want to frustrate your wireless users with inconsistent performance, and most important, you can't afford to unnecessarily strand your investments.

PAGING NETWORKS

The primary nonvoice wireless application is sending and receiving text messages via a pager. Paging still has distinct advantages today. The paging networks cover 90% of the United States, almost guaranteeing your message will be delivered. The technology works inside buildings and in underground parking areas and is very reliable and low cost, delivering messages for about one-third the cost of a telephony network.

Pagers have evolved into a full e-mail delivery tool, with two-way messaging and the important always-on feature. The popular BlackBerry device and others like it lead this segment, with others joining quickly. Many businesses have adopted standard paging networks instead of the telecommunications networks for mobile e-mail. The popularity of the BlackBerry device supports the notion that simple e-mail management can be done efficiently and relatively cheaply. The downside to paging is the inability to run application software and the difficulty of integrating pagers with the enterprise for more than e-mail.

Text messaging via mobile phones is catching on as well. The truncated or short messaging services (SMS) allow a short (50 character or less) message to be sent over mobile phones for minimal cost. The short messages are appropriate for the small screens on mobile phones. These messages can be sent and received on an always-on basis, so they are valuable for sending alerts or specific information that is expected by a user. Short messaging is a poor man's e-mail, but it works; however, e-mail is the nonvoice message of choice. As a result, do not expect SMS to be widely used by business until the two are converged in some manner.

MOBILE DEVICES

All of the effort and investment made in upgrading the networks is meant to enable faster and more reliable movement of data, but you must have a device that can process and display the data. There are more than a few devices to choose from when going mobile, and device manufacturers are in a frenetic race to increase usability and processing power. The main classes of devices that are capable of data processing are the 3G-rated phones, wireless-enabled PDAs or the more powerful pocket PCs, and a device that combines a phone and a PDA called a smart phone. There are also wireless-enabled laptops, clipboard or tablet-styled PCs, and a variety of customized, rugged mobile computing devices similar to what you see the UPS driver holding when he or she drops off a package.

Several factors must be considered when choosing the mobile devices that will interact optimally with your specific data application. A phone may work for very simple applications, but for the more complex applications, a PDA or pocket PC may be required. The factors to consider include the data entry methodology, screen size and readability, processing power, and battery life. The point to remember is that although there are many choices of devices—and many of the new middleware software packages work with many different types of devices—you have to consider the end-to-end experience of your wireless deployment, which of course and most important includes the user's device!

EXTENDING THE ENTERPRISE SYSTEMS TO WIRELESS

There are three basic blocks of work involved in extending your corporate enterprise system to wireless. They are discussed in greater detail later in the book and are included here in summary form to begin orienting you to the big picture of corporate wireless. The three blocks are:

1. Enabling corporate data for use on mobile devices
2. Installing the middleware to facilitate synchronization and movement of data back and forth from the communications protocols to your corporate network protocols and databases
3. Developing applications for the mobile user on your chosen devices and networks

The process of enabling desired corporate data to be displayed on mobile devices ranges from a straightforward scraping of data from your corporate Web site to a more complex approach of selecting specific enterprise databases and interactive corporate applications to be repackaged and extended to a mobile device. Customized programming is almost always required.

The second block involves using software called middleware that facilitates the use of mobile devices with the corporate system. Middleware works to synchronize the transfer of data back and forth from the mobile devices to your corporate network protocols and databases. Several companies (including Microsoft) offer off-the-shelf middleware solutions that have preprogrammed links to various mobile devices. This still requires customizing the linkage to the corporate business systems and databases.

The last block of work to consider is developing applications for the mobile user on your chosen devices and networks. If your mobile user application is very simple, it may already exist in a preprogrammed format; however, for most businesses, programming is required to mimic ordering forms, product specifications, or other company-specific information fields.

These three blocks of work will take at least six months to complete and cost between $250,000 to $1 million for most companies, and even more if you are a larger enterprise and still more if your applications are more complex or real time. Two chapters in this book are devoted to discussing the methodologies and project path to accomplishing the extension of complex corporate applications to mobile users.

WIRELESS LOCAL AREA NETWORKS

An important subset of the wireless world is the wireless local area network (WLAN). This is wireless without the telecommunications networks. Imagine all of the PCs and printers in your office without cables. Wireless local area networks provide traditional local area network (LAN) connectivity, speed, and reliability over short distances usually inside an office building, warehouse, or on a campus. A wireless LAN consists of one or more mounted antennas that receive and transmit data to laptops and PCs equipped with special cards that are roughly the shape and size of a graham cracker. Users of WLANs are able to roam freely within the coverage area while maintaining their network connection. Wireless LANs are becoming increasingly popular as a way to untether workers while maintaining high-speed connectivity. Wireless LANs are found on numerous college campuses, in most large airports, and in Starbucks coffee shops.

The name of the most popular standard (communication specification) for WLANs is 802.11b. The confirmed commercial range is around 300 feet, and installing transmitter/receiver nodes (like an antenna) around a building provides overlap and constant coverage. Wireless LAN technology is getting more affordable every day as more providers enter the market. Areas where employees move around frequently, such as warehouses and retail stores, are excellent candidates for WLANs. The specification for a higher-speed WLAN is 802.11a, which has more range in addition to higher speeds.

Consider the Internet and all enterprise computing as water in a huge pond. Wireless LAN access points are like lily pads on the pond. Mobile users are the frogs that jump from lily pad to lily pad. Stems from each lily pad extend into the water. Soon lily pads will cover the entire United States and eventually, a large part of the civilized world. The lily pad analogy sounds trite, but it was expounded by Alessandro Ovi, technology advisor to the European Commission. Wireless LAN lily pads are growing at a substantial rate, with more than 26.5 million units expected to be shipped in 2003, according to Gartner, Inc., the technology research firm. What is really exciting is that newer wireless devices are multimode, allowing connection to the high-speed 3G networks and WLANs. Look for a lot more to emerge with this technology in terms of higher speeds, better quality, increased security, and lower costs: WLANs are a powerful tool.

Another type of WLAN is called Bluetooth. It is very visible in terms of brand name, yet somewhat disappointing in practical widespread use. Bluetooth envisions a wireless personal area network (PAN), where users can be wirelessly connected to the various devices (e.g., phones, laptops, printers, headphones) in their personal circle or bubble of connectivity. The Bluetooth chip, which enables the communication, must be embedded in each device. Bluetooth can penetrate walls and windows, but it only has a range of 30 to 90 feet. Obtaining real value from some forms of Bluetooth communication is a practical stretch. For example, with Bluetooth you could send a file to your printer from your PC without the printer cable, if both devices were fitted with the Bluetooth chip. An amazing breakthrough! Hundreds of dollars in new technology to replace an inexpensive cable. If I take out the sarcasm, there may be value, but it is based on everyone using Bluetooth everywhere and doing it better and cheaper than WLANs.

REVIEW OF THE PLAYERS

There are many players to keep track of in the wireless space. I mention the names only to help you categorize them and so you can recall them when you see announcements in the future.

- ○ *Networks.* Competition seems to be alive and well as the major telecommunications providers fight over dominance in the mobile data space. AT&T, T-Mobile (formerly VoiceStream), Verizon, Cingular, Sprint PCS, and Nextel are the major providers. Each have plans underway to complete migration from their 2G networks toward some version of 3G. For businesses planning wireless deployment, the three primary considerations in choosing a network provider are coverage, reliability, and cost.
- ○ *Handheld device manufacturers.* The players in this space are categorized by the class of device they produce. The major phone manufacturers are Nokia, Ericsson, and Motorola. In the PDA and pocket PC category, Palm, Visor, Compaq, HP, Treo, and Microsoft are the primaries. In the smart phone category, consider Kyocera, Samsung, Handspring, and Microsoft with its Stinger Smartphone.

In the tablet PC category, you have Microsoft and Fujitsu. There are other second-tier players in every category and plenty of choices.

○ *Handheld operating systems.* This describes the software that runs the wireless device and supports application programs on the device. For mobile phones, it is more of a display option than an operating system. For wireless computing using a PDA or pocket PC, your options for off-the-shelf operating systems are really only two: the Palm OS and the Microsoft Pocket PC. There are others, but when considering the integration required with your corporate systems and the forward compatibility you want, you should carefully consider if one of these two options will work before seeking others.

○ *Middleware providers.* The middleware provider is an important choice. Chapter 7 includes a detailed discussion on middleware with checklists and possible failure points. You would select the middleware provider after you have developed your strategy and goals. You will need your IT teams to conduct a thorough analysis overlaying the middleware functions with the entire scope of your wireless deployment. There are dozens of good middleware providers; two of the majors are Microsoft and Synchrologic.

○ *Consultants.* A growing number of consultants are focusing on the wireless market, including the major accounting firm specialty units. Their focus is to assist companies with the challenge of taking their enterprise wireless. A project as complex as this one is worthy of a consultant's fee, but only if the consultant has broad practical experience that can save you time and money. Beware of consultants who begin their pitch with reinventing your enterprise. Because a sound wireless strategy supports specific business processes and ultimately creates a cost advantage, the consulting firms that have a holistic view of business would be my favorite. The major accounting firms have management consulting groups with subspecialties such as wireless. Working with them can be beneficial if you narrow the scope and make sure you get a top team with experience, not the rookies selling off the track record of a few specialists who are not deployed for you. *Tip*: Seek out strong regional firms with wireless as a specialty.

REFLECTING ON THE CHAPTER

Wireless technology is moving from the pilot mode into the practical. New generations of high-throughput networks are deployed and expanding. New and more powerful devices are being offered, and more off-the-shelf software tools to help the enterprise deploy wireless are available. Planning the wireless application today for tomorrow's use includes careful consideration of the present and a bit of thoughtful hedging about the future.

2

Wireless Workforce

The opportunity to create a wireless workforce is the competitive differentiator of the decade. According to Yankee Group, a Boston-based technology research firm, approximately one-third of the U.S. workforce, or about 43 million people, were mobile workers in early 2002. Business executives believe this group of employees, if wirelessly connected to the organization, could achieve a 12% to 16% improvement in productivity or equivalent cost savings. The potential of growing the bottom line by improving worker productivity with wireless is worth the time it takes to investigate and may also prove worthy of your investment.

When should you consider wireless for your workforce? Wireless makes sense for workers whose location changes frequently, for work processes where information is compiled and recorded more than once, and whenever timely information would allow a process to be completed faster. An additional application may exist in industries that require an audit trail for compliance purposes.

Moving your workforce to wireless is more than submitting expense reports over a PDA, which may indeed be convenient but does not necessarily bolster the bottom line. The true benefit of wireless for the workforce is realized when business processes, customer relationships, or revenues are optimized or enhanced. Let's examine several jobs and industries that could capture real benefits from wireless applications, beginning with the salesforce.

WIRELESS FOR YOUR SALESFORCE

In good times or bad, every company is looking for ways to grow its top line. The favorite way to accomplish that goal is to boost the productivity of the salesforce. For years, companies have armed their salespeople with promising new tools, such as pagers and cell phones, to improve their individual productivity and to tighten their connections with their customers and their internal support organizations.

Today, selling is a high-touch process supported by high tech. Those who thought the Internet and e-business would replace the traditional salesperson were mistaken. A 2001 survey of 200 U.S. companies by the research firm ActivMedia revealed that just 8% of firms offering online sales were actually closing a majority of sales through the online channel. That means 92% were closing orders using their salespeople, often using the Internet in a support role. Notwithstanding the small but steady yearly increases in online sales, the salesperson is alive and well, but something has changed.

The Internet has raised customers' expectations to a point where they demand immediate availability of information about pricing, inventory, shipment status, and pending quotations. As this kind of transactional information becomes increasingly available and expected by customers, salespeople must have all of the information they need—anytime, anywhere— to know of any problems before their customer knows of them.

Wireless enables a salesforce to stay tightly connected to its company's information stream while staying mobile to serve customers. When in the field, salespeople can initiate queries and take action to develop on-the-fly solutions to real-time problems. The key is providing salespeople with the right information at the right time, not all of the information all of the time. For many salespeople, the right time is when they are on the road, between customer calls, or just before or even during an important client meeting.

Consider Producers Lloyds Insurance, a leading crop insurer that has field agents who need precise pricing information for dynamic pricing scenarios. Producers' agents are literally in a farmer's field much of the time, working with customers to assess, price, and provide insurance coverage for crops in various stages of development. Producers' agents access their corporate database wirelessly to formulate pricing that is based on crop type, timing in the season, weather predictions, and the coverage level required. Being able to access the corporate data and develop a detailed price quote in real-time is critical to making a commitment and closing the deal on the spot.

Question: If two companies were competing for your crop coverage insurance business, and one provided you with a firm quote on the spot, whereas another company's agent said "Let me get back to you with the quote," with which one would you want to do business?

It wouldn't be fair to talk about the salesforce without addressing customer relationship management (CRM). Today, CRM programs are software and processes that aim to create an integrated information model for planning, scheduling, and controlling all presale, selling, and postsale activities in the organization. Companies have taken a variety of approaches to CRM, including buying expensive software and paying even more expensive consultants to help the companies figure out how to use it. Granted, consultants are often able to help companies focus, organize, and execute more efficiently and with more precision; however, most companies already have in place many of the informational components needed to extend basic CRM capabilities to their salesforce. Combined with a strategy and some tools, wireless can be an enabler to form a cohesive, effective CRM program.

Most professional salespeople would agree that selling is still very personal. If you don't have trust and a relationship with the customer, forget it; however, your salespeople compete with competitors' salespeople who also have relationships and the trust of your target customers. With all other relational and performance issues being equal, the edge goes to those salespeople who can deliver accurate information to their customers in the timeliest manner. This is nothing new. Customers have always required timely information on their orders; the difference is that today their expectations have increased to match the improvement in the amount and availability of information. Customers expect to know more—sooner.

Today's salespeople must communicate constantly with their customers and with their organizations, even though they are on the road or in the field for a large part of the workday. Regardless of their whereabouts, salespeople are expected to be always available and always informed of each customer's situation. Responsiveness is key to growing the business. How has the standard for information and organizational connectivity in selling changed over time to today?

Exhibit 2.1 illustrates the evolution of the mobile salesperson.

Exhibit 2.1 Evolution of Salesforce Automation

Personal Organization		Static Organization Link		The Real-Time Sales Organization
Address book	⇨	Contact management files	⇨	Live notes and info sharing
Calendar		Calendar management		Calendar synchronized & updated
To-do-lists		AM/PM e-mail		Always-on e-mail + alerts
		Pricing models		Actual pricing
		Quotation generation		Quotes synched with the business system
		Product configurators		On-hand inventory info
		Order history		Order status
		Request for information		Sales leads e-mail in real time
		Sales leads called		

Stand-alone PDA ⇨ *PDA Docked & Downloaded* ⇨ *Wireless PDA—Always-on & Synched with the company business systems*

Where the customer interaction processes are most basic, salespeople need simple personal productivity tools such as a stand-alone PDA and a cell phone. In order to add value in faster-paced sales scenarios, salespeople must make more information available to the customer as well as be more proactive and involved in the details of managing the account. Downloading customer and corporate information once a day will empower the salesperson to a degree. For many salespeople, pricing and quotations support doesn't require real-time feeds, and a simple phone call can still be made to ascertain up-to-the-minute lead times and inventory availability.

The next highest standard to achieve is empowering the always-available, always-connected salesperson. Here, salesforce requirements include immediate and continuous access to e-mail and contacts and calendar management, always-on receipt of customer alerts (e.g., credit problems), and CRM capabilities, such as access to current order status and pricing. A typical solution today is a PDA and modem combination equipped with a CRM application synchronized with corporate systems and a Web-enabled phone for voice and short e-mail/customer alerts. An even better alternative may be a smart phone integrating both PDA and telephony capabilities.

Customers will want to work with salespeople who are always connected and available because they can count on them to be always informed about the specifics and details of their accounts. These salespeople are also able to be proactive about informing their customers and solving problems. Isn't this the kind of salesperson you would want calling on your company? Doesn't it describe the kind of salesperson you want to have calling on your customers?

If the nature of your business is slower paced, where product and order information changes rarely, pricing is consistent and semifixed, and lead times are always accurate, then you can moderate the real-time and synchronized elements downward; however, you could also change the game in your industry by raising the bar and differentiating your company. If you think your business could benefit by evolving to real-time customer service, you need wireless and a little process reengineering!

Consider Dallas-based chemicals manufacturer Celanese. In the past, when customers would call Celanese salespeople, they would politely excuse themselves from the call and then quickly call the home office customer service personnel who had network access. After retrieving the required information on products or orders, customer service would call back the salesperson and relay the information for hurried notation by the salesperson. Only then would the salesperson finally be able to call back the customer with an answer. The process took anywhere from four hours to a full day. Guess who would get the hot orders on products where delivery and price were equal? Not Celanese.

But wireless changed all of that for Celanese. Now salespeople use mobile handhelds linked to a Web-based server to access the company's enterprise system and retrieve information in real time. Salespeople can answer customer questions, on the spot, demonstrating Celanese's commitment to excellent customer service. Celanese is ringing up more orders, and more important, it anticipates an increase in customer satisfaction and subsequent improvements in customer retention and penetration.

Is wireless right for your salesforce? Below is a list of some of the capabilities that wireless makes possible in a sales situation. See if any of these make sense in your business:

○ Before calling on clients, salespeople can check on outstanding orders, enabling them to be fully prepared and up-to-date on the clients' accounts. Knowing in advance about delays or problems allows salespeople to be proactive and to give customers confidence that they are on top of the situation.

○ Salespeople also can submit changes to an order, plus make inquiries in real time right from the customer site.

○ If customers wish to place an order, salespeople can pull up the product specs and pricing, access inventory availability, and share it with customers to make the right decision in ordering.

○ Requests for quotes can be processed from the field, as salespeople receive them.

○ Between calls, salespeople can record information about each sales call, enabling more accurate and timely call reporting and sales management.

○ Salespeople can receive alerts while on the road, notifying them of a customer crisis brewing. They can proactively contact the customer with solutions instead of waiting for an angry call from the customer.

○ Being able to stay connected on the road allows salespeople to schedule their time more efficiently and effectively, keeping them proactive and responsive.

Another good example of wireless enhancing salesforce effectiveness is at MGI Pharma. Pharmaceutical salespeople make numerous sales calls each day to dozens of doctors, giving away trial drug samples and explaining the benefits and side effects of new drugs. Providing the samples entails obtaining a signature from the receiving doctors and each salesperson is required to take notes on the doctors' questions and responses to the new drug. MGI Pharma's salespeople used to jot down notes and accumulate drug receipts from each visit, which they would hold and then organize and input into the company's computer system once they returned to the office every few days.

Now, armed with their new wireless application, salespeople can record the information from each call as it is completed, eliminating the compiling of paper and rekeying of data back at the office. Salespeople can also have the doctors sign directly on the PDA, which records their signature digitally for the required recordkeeping. The technology keeps the MGI salespeople on top of their accounts because the information is captured while it is fresh on their minds. Management also has a better handle on the numbers and quality of sales calls per day.

Do your salespeople have laptops, but your company isn't getting the productivity it anticipated? You may want to switch to handhelds, which some salespeople will find easier to use. After introducing handhelds to augment laptops, Alcatel, a telecommunications equipment maker, reported a fivefold increase in the amount of data sent by field salespeople.

In addition to greater utility, handhelds offer a lower total cost of ownership than a laptop (examined later in this chapter). The bottom line with wireless for the salesforce is greater efficiency and better sales management, but it may also mean the difference between keeping and losing your customers.

WIRELESS FOR YOUR SERVICE TECHNICIANS

A study by D.F. Blumberg Associates, Inc., in 2001 revealed that approximately 500,000 organizations provide some form of field service, employing around 2.3 million field service workers.[1] The economics of supporting investments in vehicles, tools, rolling inventory, and training requires the maximization of billable work, but all too often, technicians end up returning to an office to re-create or enter data from their service calls, losing valuable service time in the process. Given this situation, it's not surprising that some of the greatest benefits of early wireless applications have been achieved by companies providing mobile repair workers with wireless access to information and transaction processing capabilities. When equipped with wireless computing tools, worker productivity per call improved, fueling an increase in the number of customer calls per day.

Mobile service technicians move frequently between customer sites with myriad tasks to perform. Almost every service call follows these eight steps:

1. Receive a customer call for service.

2. Determine availability and contact, and dispatch service worker.

3. Travel to a customer's work site.

4. Assess parts required.

5. Search for and acquire parts.

6. Complete the work, close the work order, and prepare customer billing.

7. Move to the next customer site.

8. Report daily time per job.

These steps are for straightforward service calls, and each step can be positively enhanced by wireless; even the driving step can benefit from automated directions and traffic avoidance! Using wireless and mobile computing technology, many of the steps can be significantly streamlined. Technicians can accept or decline assignments according to how their current service calls are proceeding, schedule a return visit, receive information on the next call, and/or prepare a bill for review and digital signature by customers, thus speeding the billing process. Technicians can stay in touch, instantly exchanging service schedules and route updates.

Reduction in paper processing by entering data only once at the customer site and transmitting the billing information to a central system can save labor costs and speed billing. An additional benefit of prompt, more precise recordkeeping on service calls (time spent and parts used) can help avoid the typical 20% to 30% rejection rate of service invoices by customers who object to the charges. Much of the needed information can be entered offline if a connection is unavailable and then later synchronized online with the information from the technician's handheld device and automatically uploaded to the company's central database.

Because approximately 70% of field service calls involve ordering parts, connecting the service technician to the corporate inventory and ordering database can enable faster inventory searches, part allocation, and ordering. More complex work may require the technician to access online installation or technical manuals, review the previous call history, order nonstocked parts, and inform the customer about the price and delivery of parts.

Wireless, together with supporting business applications, can significantly improve the scheduling, worker productivity, inventory management, and overall customer satisfaction with service operations. Customers will be happier. Sales will increase. The bottom line will improve.

For example, early one morning while I was sitting at my kitchen table writing this book, a Sears repairman named Stan arrived to fix a broken rack roller and find out why the water wasn't draining from my dishwasher. He walked in with a rugged laptop sprouting a little antenna. The unit was slung over his shoulder and had a microprinter tucked into a pouch on the strap. He set up the laptop on my kitchen counter and called up my work order information, which had been downloaded into his unit early that morning from his home just before he started his workday.

After assessing and curing the drain problem, Stan proceeded to search for the replacement roller part. If it was a Sears part, he could use his laptop to search the Sears inventory database. My dishwasher is not a Sears brand, so he had to call a special toll-free number and have a Sears customer service agent help him with the cross-reference equivalent of Sears parts. Stan said it would be great if he had cross-reference capability on his laptop as well as a link to actual lead times, but for now the phone call was required. Once he had the cross reference to a Sears part number, Stan entered the data into his laptop and first checked if the part was on his truck, which it wasn't, but it could have been. I asked him how many parts he had on his truck. I was told he had hundreds of parts and the computer helped him track what was on his truck. For my part, it would have to be ordered. He was then able to pull down the price and lead time, and on my approval, he added the part to my service order.

Stan said that at night, he takes the laptop home and plugs it into his phone line. The unit is polled by Sears during the night and the upload is completed—just in case he was out of range during the day on certain calls. He also receives a download with his call schedule for the next day, which he reviews and confirms in the morning before he starts his service schedule. During the day, he also can receive new service calls via wireless, if he is in range. Sears uses a proprietary radio network that has great coverage in metropolitan areas but is less reliable in remote areas.

I wanted to pay for the call and part on my Sears charge, but he could not verify my account number or available credit through his laptop application. He would have to call the tollfree number again, but decided he trusted me. He allowed me to give him my Sears account number, and he entered it in the unit. Stan printed my receipt, which I signed. He indicated the printer was "a little wimpy," and the customer copy came out so light that it was unreadable. Thus his standard practice was to print the receipt twice, giving me an original copy. The receipt was very complete, with all of the details on costs. I was very pleased with the precision of the call.

I asked Stan if he went in to a main office at all, and he indicated that only recently had the entire service department begun leaving directly from their homes to their first service call. This saves thousands of hours in travel time. I asked him why they had only begun to do this recently. Hadn't they had the wireless units for a few years? Yes, he replied, the wireless units have been around for a few years, but Sears recently made an important process change.

In the past, Sears would inventory thousands of replacement parts for all types of units in its local warehouses. The service technicians like Stan would drive to the main office and warehouse every morning to pick up the parts that had been previously ordered, store them away on their trucks, and wait for central dispatch to schedule them to go back to a customer's home to install the new parts.

Recently, Sears decided to eliminate the warehouses completely, along with the trips back and forth to the warehouses by the service techs. Parts ordered are now delivered directly to the consumer's home by UPS. When the consumer receives the parts, a note on the box prompts them to call a Sears toll-free service number, reporting the arrival of the parts and scheduling the follow-up service call. Many other companies could take a page from the Sears playbook. According to the study by D.F. Blumberg Associates, Inc., a research firm based in Fort Washington, Pennsylvania, inventory costs in service firms could be reduced by 20% to 25% using wireless to search databases in real time combined with improved inventory management software.

Stan said that many parts are ordered from the factory anyway and that the overall time delay for the consumer is about the same. I felt good because I knew my special parts were already ordered, and I had an estimated date for Stan to come back to install the new parts (Stan had a calendar on his laptop unit where he could show me the possible dates he could return and see which ones might be good for me.) He keyed in my preferred date to his laptop as a reminder to himself, subject of course to the parts arriving. Stan also gave me some discount coupons good for my next service call, plus $5 off my next apparel or jewelry purchase from Sears.

As Stan pulled away to his next service call, I thought about how Sears had built continuous process improvements around the wireless system it had introduced eight years earlier. Sears had eliminated the working capital tied up in inventory parked in warehouses; saved thousands of hours of drive time for some 13,000 service technicians; boosted its profit margin by adding a shipping and handling charge, and at the same time, it got me to help them schedule my follow-up service call. The best part is—I liked it!

Somebody at Sears is thinking—not exclusively about wireless per se, but about continuous business process optimization. Wireless is just a tool to be used to maximize operational efficiency and customer satisfaction, both leading to a positive impact on the bottom line.

The benefits could apply to almost any service organization. According to the Blumberg study, using wireless applications in field service organizations can improve technician productivity by 15% to 25%.

GE Medical Systems offers a great example of wireless deployment in a service workforce. GE offers extended warranty and service on sophisticated medical scanning equipment. Delivering fast service and information updates to customers is critical. GE calculated that its 2,400 service technicians spent as much as one-third of their time on administrative tasks, such as schedule updating and call reporting. These tasks were often delayed for days, causing customer reporting to be inaccurate. It all reflected poorly on GE, so the company's goal was to reduce administrative time to 5% and increase the timeliness of data from one week to four hours.[2]

GE's service technicians carried laptops, but booting them up took too long. They switched to smart phones, which combine a digital phone with a Palm Pilot. The GE Medical Systems Web site was modified to work with the Palm devices for inventory queries and checking on parts deliveries through FedEx. Now field technicians can get online faster and update customer records more often. According to GE, the reaction from technicians has been very positive because the tool is more usable. GE was able to increase the productivity of its field service technicians, improve the timeliness of data reporting, and enhance updating of equipment status, which increased customer satisfaction. GE did all of this while lowering its cost of hardware (switching from laptops to smart phones) and lowering the central support required by field technicians.

The ROI and customer impact of wireless for the service technician workforce is so powerful that we can look for this segment to outpace all others in wireless utilization. In the near future, technicians will perform even more tasks wirelessly and remotely. In one scenario, more experienced technicians could see and diagnose a problem remotely using digital microcameras, and would then instruct a less qualified (and less expensive) junior technician on how to repair the special control panel or other complex piece of equipment. Diagnostics could be uploaded and analyzed, providing information on the best course of action.

From the initial dispatch call to ordering parts and closing out the ticket for billing, every step in the field service process can be enhanced with wireless. Just as Sears did eight years ago, you need to get started with wireless today to prepare your service technicians and business operations for the future.

WIRELESS FOR WORKERS ON THE MOVE

Tens of millions of workers other than sales and service employees also can benefit from wireless technology. Executives, lawyers, accountants, management consultants, and engineering project managers all travel extensively, meeting with clients, vendors, and each other along the way. To be effective, they need immediate and continuous access to e-mail as well as remote electronic access to contacts and calendar management, which must be capable of synchronization with the primary business system. In addition, architects, consultants, and others need superior retrieval and file management of graphics-intensive documents. The solutions that are available today include a smart phone for telephony, e-mail and contacts, and calendar management. Add to that a laptop containing an 802.11 network card, which allows users to access a wireless network while in airports, convention centers, hotel rooms, and other venues on the road. The phone can be synchronized with the laptop for contacts and calendar management, and it can also be used to provide digital modem capabilities if an 802.11 network is unavailable.

When e-mail is the primary requirement, a good alternative is Research in Motion's (RIM) BlackBerry device and equivalents. These devices are always on and can be coordinated with primary business systems to send alerts on specific e-mails. Not only is the BlackBerry very functional, but it is also cheaper to own and operate than a laptop and much more efficient to use in accessing e-mail. Many other PDAs have integrated the functionality of the RIM BlackBerry for fast management of e-mail; there are lots of choices.

According to figures from the Boston Consulting Group, the BlackBerry costs $900 per year to own and operate. Comparing the cost to a laptop, for an employee earning $100,000 per year, the system pays for itself in less than a year by saving 10 minutes per day in setup, dial-up, and download time, not including the value of getting vital information faster. In addition, incidental support costs for a BlackBerry are lower than for a laptop. The BlackBerry's corporate system can be set up in a matter of hours; a simple interface is added to the link between the company's e-mail server and the wireless operator. One downside is that some employees who have been using laptops will want to keep them plus have the BlackBerry, so there are no real avoided costs, only improvement in communication, but time is always money.

Another viable application for wireless is supporting compliance, quality control, and other business requirements. Wireless can be deployed to boost the efficiency of field auditors in the course of doing their highly mobile jobs. Domino's Pizza, for example, has a staff of more than 50 quality control auditors who visit hundreds of franchises across the country to conduct audits and surveys. Before the auditors go out to the field, they synchronize with corporate systems and load their Palm devices with surveys and specific information about each franchise that they will visit that day. After completing the survey, the store manager can sign the auditor's report right on the Palm. Using a wireless modem attached to the Palm, the auditor sends the information back to the Domino corporate system, uploading the report and then downloading the details of the next franchise to audit.

ServiceMaster, a nationwide service company, utilizes Palm devices at 20 Greyhound bus terminals to record the results of bus inspections. The company shelved its paper-based system and issued Palms to its inspectors, who now file their reports daily instead of weekly. Inspectors enter data into their Palms and then upload to the ServiceMaster server, which compiles the data and produces reports for Greyhound. The cycle time for reporting to Greyhound was reduced from 10 days to two. This allows Greyhound to correct problems faster because it had the information sooner.

At Volvo, which imports more than 145,000 cars per year into the United States through four ports, the process of collecting and consolidating inspection data is a daunting task. Inspectors have switched from paper to Palms. Inspectors now record information on the 40 items they are required to verify on each new car by following a checklist on their Palm devices. Using wireless devices has reduced the turnaround time for inspection data from weeks to one day. Irregularities can be identified earlier in the cycle, enabling dealers to fix cars and get them on the showroom floor sooner.

Wireless has even invaded upscale restaurants. At Republic restaurant in New York City, servers take orders on their handhelds and then beam the orders to the kitchen via infrared or radio frequency. By avoiding the trip to the kitchen to turn in the order, waiters spend more time with their customers and attend to more tables. This system has effectively lowered the amount of staff needed. In addition, management reports that turnover time of tables has decreased from approximately 50 to 35 minutes, a solid improvement. In addition to automating order entry, the new system provides easy reference to a changing menu and pricing. It also improves service by reducing wasted trips and traffic in and out of the kitchen. Waiters are always visible to their customers. They also can process credit card transactions right at the table, get the customer's signature, and print a receipt. They can even message the valet to bring around the patron's car!

New-car shoppers are using the Internet to research and obtain quotes from dealers. Responding quickly to an Internet request can be a positive start to a relationship that ends in a sale. Jim Hudson Lexus/Saab was receiving more and more Internet requests for quotes and quickly learned that the first dealer to respond to the consumer had a better chance of building the personal relationship required to close the sale. At Jim Hudson Lexus/Saab, one sales manager at each location is equipped with a BlackBerry wireless e-mail pager to monitor information requests from Internet customers. The new process has reduced the response time from between one and six hours to a new benchmark of 30 minutes. Deals that close have increased from 10% to 15%. Internet Sales Manager Tony Albanese, who also carries a BlackBerry, says it helps ensure that he can respond to a customer's inquiry while that customer is logged on the Web. Based on the competitive advantage provided by the new process, Lexus now mandates that dealers must respond to customer e-mails within minutes, not hours. Wireless has helped set a new benchmark for customer service.

Are there more workforce niches where wireless could work to improve processes? Definitely. Look for additional applications to surface as we move to 3G speeds.

My favorite example of applying wireless is in a unique job position—the Landing Signal Officer (LSO) on U.S. Navy aircraft carriers. As Navy jets make their way in to land on the carrier, the LSO's job is to grade pilots on their steadiness, accuracy, speed, and other factors, during both day and night landings. These grades are critical to helping pilots improve their performance and to maintaining high safety standards in the high-risk environment.

The LSOs are placed in a tiny space near the front of the flight deck at the edge of the ship—right where the aircraft first touch down. They are subjected to constant wind and occasional water whipping through their work area, first from the ocean and then from the jet exhaust. They work in daylight and in dark. Previously, the LSOs used a piece of paper on a clipboard to write down the grades immediately after each landing. They would then forward the paper for input into the ship's flight-grading computer system. The pilot debriefings were often over and many hours had elapsed before the grades were posted. The environment wasn't the best for a fixed large terminal, but improvement was required.

Now, the LSOs use Palms and a stylus to record the grade for each skill measured *as the planes land.* After entering the information about each landing, the LSO synchronizes the Palm with the shipboard network. The information is then made available for everyone to review, immediately following the landings. Pilots see their grades as soon as they arrive from their planes into the debriefing room, and the double entry into the ship's computer system was eliminated.

REFLECTING ON THE CHAPTER

Taken together, these examples clearly indicate that wireless communications and computing for the mobile workforce is the productivity super-charger of the next decade and the weapon of choice for automating selected business processes across multiple industries.

Wireless also spans multiple job functions, ranging from salespeople and insurance adjusters to delivery drivers and restaurant servers. Will it work in your industry with your workforce? I believe wireless can benefit many situations where exchange of information between mobile workers and the corporate business system would improve a process or simply empower an employee to provide better customer service. You just have to look for it!

There is more to making wireless work than just a device or software, however. It requires taking a holistic approach to the particular business process targeted for wireless transformation. Once you have zeroed in on the target process, the challenge will be extending the enterprise system to work efficiently with the new wireless process as well as making some important decisions on devices and communications. Ultimately, it may take some process changes as well.

Is it worth it? Quite a few companies believe so and are already moving full steam ahead.

ENDNOTES

1. Study from D.F. Blumberg Associates, Inc., as reported in an online whitepaper published by Arcstream, "Field Service Opportunities."
   Access at *www.arcstreamsolutions.com,* or call 617-393-2400.
2. Howard Baldwin, "2001 Mobile Master Awards for Enterprise Deployments," *mbusiness* (January 2002).

3

Wireless Warehousing and Logistics

Wireless is transforming warehouse and logistics operations in highly visible and measurable ways, including the following:

○ Reducing errors in items picked and shipped

○ Increasing worker productivity, measurable in units picked per hour

○ Giving salespeople and customers end-to-end visibility of their orders

○ Optimizing fleet deployment and deliveries

○ Compressing delivery cycle times

In short, the benefits of deploying a wireless warehousing and/or logistics system are significant, and they cascade throughout your supply chain.

Maximizing inventory turns while minimizing stockouts and errors are the primary business objectives of any warehousing operation. Tracking goods from the time they are received and put away, then picked, packed, and shipped out again is a paper-intensive and error-prone process. The warehouse industry estimates that warehouse errors result in the shipment of the wrong product as much as 5% of the time. Errors produce a ripple effect on costs, such as those for processing return and reshipping, which together average about $100 per shipment. Repeating work steps also means disappointed customers and productivity losses.

Integrating wireless and barcoding into inventory management eliminates almost all of the errors, and according to Larry Gordon, a Research Fellow for Jupiter Research, it provides "close to perfect" information.[1] Let's examine the key elements a little deeper.

The wireless warehouse and logistics environment comprises three basic components: the wireless local area network (WLAN), the handheld processing unit equipped with barcode scanner, and the application software package and adapters necessary to process the data and integrate it with the business system. Let's review each of the components and some examples of various applications.

WLAN TECHNOLOGY

Most midsize and larger companies have a corporate local area network (LAN), which is hardwired to workstations throughout the business. A basic 802.11b WLAN provides LAN connectivity, speed, and reliability over distances that range between 200 and 300 feet with no wires.

A wireless LAN consists of a mounted antenna that transmits data to laptops and PCs equipped with special cards that cost between $100 to $200 and are about the size and shape of a graham cracker. In warehousing, data are usually transmitted to specially equipped wireless handheld units that are used by warehouse workers. Workers are able to roam freely within the coverage area, while maintaining their high-speed network connection.

Let's quickly review the technology background of WLANs. The most popular WLAN is known as 802.11b. The numbers refer to the communication standard established for any WLANs to make them off-the-shelf and compatible with each other. The confirmed commercial range of 802.11b is about 300 feet. By installing transmitter/receiver access nodes (e.g., an antenna) around a building, you can cover a large area. The access point converts the wireless data into wired data and vice versa, acting as a bridge point between the wireless user and the wired LAN. With a little overlap, a WLAN can provide constant coverage over a large area. At around $500 per node, the 802.11b technology is not that expensive and it is coming down in price as more vendors enter the market. One access point can serve 50 to 70 users.

Moving outside of the warehouse and into the office, the WLAN is also a viable alternative to a hardwired network. The total cost of ownership of a hardwired network is about the same as a wireless network. Hardwired network costs include both the initial wiring costs and the ongoing costs of constant moves, additions, and changes that a corporate network endures. To make the same network wireless, workstations would be fitted with the transmitter/receiver module and could then be moved around the office as needed without a hardwired network connection. Yes, a return on investment is possible with wireless LANs!

This helps explain why wireless LANs are increasingly popular as a way to untether workers while retaining their high-speed connectivity. Frost and Sullivan reported in late 2001 that annual shipments of WLANs will reach more than 20 million units by 2007.[2] These units are primarily for private use, but public WLAN sites are growing as well. The number of 802.11 hotspots in public venues is predicted to grow to 41,000 "nodes" and reach more than 21 million users by 2007 (up from about 600,000 in 2002), according to Analysis Consulting, a British research firm.[3] Wireless LANs can now be found on college campuses, in retail stores, and even in Starbucks coffee shops. Nonwarehouse uses of WLANs are discussed further in Chapter 10, Looking Ahead.

Another WLAN, 802.11a, is higher speed, operating three to five times faster than 802.11b, with double the user capacity and much less susceptibility to interference, but unfortunately almost double the cost as well. The 802.11a standard is required if you want streaming media (video), which is not necessary in a warehouse, but probably a plus in a campus environment. The overall reliability is good for both standards and may improve even more for 802.11b, and most important, costs will continue to drop on both types.

HANDHELDS AND BARCODING

The second component of a wireless warehouse and logistics system is the handheld unit. These units vary in size and function, but all should incorporate a barcode scanning capability. Warehouse workers who are equipped with wireless communications and barcode scanning technology can support each step of the logistics process, from receipt and shelving of goods to picking, packing, and shipping. There are plenty of reasonably priced portable scanners ($300 to $700), or you can purchase wearable wireless scanners that fit on the finger and are linked to a keypad strapped to the arm, keeping both arms free for warehouse personnel to manipulate the shipment. The wearable wireless devices are not cheap (about $2,500 each), but as more companies enter the market, the price should drop. You don't have to overspend on the system, though. A modest system with simple devices can yield superior results.

Dee Electronics, a $30 million electronics distributor that stocks between 40,000 and 45,000 stock keeping units (SKUs) opted to deploy a wireless LAN and barcode system to improve the picking accuracy in its warehouse. Each picker in the warehouse carries a WLAN-enabled terminal with barcode reader to scan every item as it is picked. If the wrong item is chosen, the picker knows right away because the scanner indicates the error with a beep. As items are accurately picked, the inventory system is instantly updated, and the status of the order is visible to customer service personnel and even to customers, who can see where their orders are in the Dee system. Parts are scanned a total of 11 times from receipt through putaway and picking, then shipping, resulting in a virtually error-free process. Even though Dee is relatively small, it has a big-company real-time inventory system to take to market as a competitive advantage.

Moving wireless into your warehouse requires adding to or replacing your existing warehouse management system to include the wireless and barcoding software. Literally hundreds of software companies provide off-the-shelf packages with application programming interfaces (APIs), otherwise known as "hooks," to allow rapid integration with ordering, sales, accounting, and other business systems. Some software customization is inevitable as the barcoding processing components of the system are integrated with the inventory and shipping components of the company business system. Many vendors sell a complete package that includes handheld wireless barcode scanners, the warehouse management system, integration modules, and the services to install and train workers on the system. The price depends on the scalability of the system and the number of workers and warehouses in the network.

An example at the higher end of the scale is San Francisco–based McKesson Corp., the nation's largest wholesale distributor of drugs. From aspirin to Pepto Bismol, McKesson's warehouse teams pick from tens of thousands of products per day. In an eight-hour shift, a single picker can track down and pull as many as 4,000 products. The former paper-based process used was fraught with human errors generating misplaced products, inaccurate inventory, and delays in orders. Based on data that supported an attractive return on investment (ROI), McKesson decided in 2000 to automate the process with wireless computers that pickers strap to their wrists—quite a leap for the 168-year-old company!

The wrist computers receive data over a wireless network inside each McKesson warehouse. They display detailed order information and the exact row, shelf, and bin where the product is located. The system will also route pickers through the most efficient sequence as they pick multiple orders. Pickers pull the items from the shelves and scan the product barcode with a scanner worn on their fingers. Occasionally, the system will run cycle counts while workers are picking, asking the workers "how many boxes are left in the bin?" This information is then entered on the wrist device and incorporated into the warehouse inventory system to verify on-hand quantities.

McKesson spent a total of $52 million, which included outfitting 1,300 warehouse workers with the wearable wireless devices at a cost of $2,500 each, and millions more to equip their 52 warehouses with WLAN. Tom Magill, former VP of Logistics Technologies for McKesson, declared that the system has paid for itself many times over and has saved McKesson millions.[4] The productivity numbers back up his statement and are impressive. McKesson reported an 8% gain in productivity, an 80% drop in incorrect items shipped, and a 50% drop in product shortages. Inventory accuracy has improved to 99.5%, which is close to perfect in anybody's book!

INTEGRATING WAREHOUSING AND LOGISTICS

Significantly reducing errors and improving warehouse productivity are immediate benefits of wireless warehousing, but what about extending the wireless concepts to the entire logistics chain? Wireless can be deployed to the movement of goods from the warehouse to the customer.

Ace Beverage, a Southern California beer distributor, is an example of a company that is integrating wireless into its warehousing and extending it further into its logistics chain. Ace delivers hundreds of thousands of cases of beer each year to hundreds of chain and convenience stores. Before going wireless, Ace salespeople visited each customer location to take orders, share specials, and so on. Near the end of each workday, the accumulated paper orders and notes were hand-carried back to the regional warehouses for processing. The paper orders had to be keyed into the system, before the process of pulling product, sorting loads, loading trucks, and assigning drivers could begin. Occasionally, salespeople would return late, delaying the next day's loading for their customers later into the night. Imagine dozens of salespeople all arriving late in the day with paper orders for entry, then a mad rush through the night in picking, staging, and loading trucks. Trucks needed to be rolling out the door by 5 A.M. each day.

With the new wireless system, Ace salespeople enter orders on a wireless tablet PC while on site at their customers' locations. All orders are in by 2 P.M., when the warehouse teams can start preparing orders by picking and staging them for loading onto trucks. Mike Krohn, VP of Finance and Administration, has reported that two of the major benefits of the wireless system are a reduction of 15 to 20 hours per week in worker overtime and savings in additional warehouse equipment Ace didn't have to buy.[5] A third benefit of the new system is that salespeople can complete an electronic order faster than a paper-based one, which allows a few more minutes for customer face time on each sales call and/or one or two more sales calls per day.

TRACK AND TRACE

Finding, managing, and tracking products shipped from warehouse to customer is the sweet spot of wireless. Extending barcode tracking into the shipment cycle will allow you to provide customers with real-time information on the location of every shipment.

FedEx got the ball rolling in the mid-1980s with its digital-assisted dispatch system. FedEx was able to more efficiently route pickups and deliveries, optimizing its workforce in the process. The company tied the process in with another innovation called SuperTracker, which provided for tagging individual shipment and location information back to the FedEx central system and ultimately to customers. FedEx was tracking packages in real time as early as 1986.

Today, FedEx and UPS also provide real-time display of shipping information on their Web sites, giving access to customers and everyone else along the supply chain. The next step for companies that use those services and want to report shipping status to their customers is to poll the FedEx and UPS Web sites, extract the information, and then with proper formatting, deliver the information to customers. By integrating FedEx and UPS information into your process, you benefit from fewer disputed shipments, faster proof of delivery, and improved cash flow.

Today, shipping can be outsourced to FedEx and UPS. The other option is to emulate what FedEx and UPS accomplish, but for a lot less.

Using off-the-shelf technology, you can optimize your delivery fleet, capture customer signatures, and provide real-time shipping information to your customers, regardless of your position in the supply chain. Let's explore how wireless is giving the trucking and delivery business a major makeover.

WIRELESS TRUCKING AND DELIVERY

There are literally thousands of transportation companies in the United States operating more than 20 million commercial vehicles. For delivery services, if you are not sending your package by mail, FedEx, or UPS, then you are probably using one or some combination of short-haul couriers, local delivery, less-than-truckload (LTL) deliveries, or long-haul trucking. Freight transport companies compete in one of the toughest industries imaginable, with profit margins as low as 2%. Even small gains in productivity or efficiency can make a big difference in profitability. Because 40% of freight transportation costs are attributable to fuel consumption, improving the efficiency of routing and dispatching has an immediate impact on the bottom line.

Trucking and delivery companies have been actively deploying technology to improve their operations since the late 1980s. On the road, drivers can use wireless cell phones to send and receive information about addresses, directions, and so on. Customers can receive alerts and be asked to respond when they are available for deliveries to be made. Two-way radios and cell phones are beginning to give way to more specialized and integrated wireless applications, in which communications are combined with location information processed by routing optimization software in real time. For example, Nextel has a phone that incorporates a global positioning system (GPS) feature (called @Road), which, when coupled with its two-way radio feature, allows a dispatcher to locate drivers and dispatch the closest one to the customer.

There's more: Wireless technology is available to allow drivers to record details of each delivery, obtain electronic signatures for future proof of delivery, and input any delay information, which enables dispatchers to more efficiently reroute other drivers. The three primary benefits are increasing efficiency in dispatching, improving routing and other information provided to drivers, and giving customers access to the whereabouts of all shipments.

An example of one company using this technology is Averitt Express, a Southeast-based transportation company that delivers freight to thousands of destinations ranging from Florida to Texas. Using its 92 regional facilities, Averitt moves approximately 17,000 shipments per day, representing 20 million pounds of freight. The process of tracking cargo, determining destinations, and scheduling was done manually with a log system and lots of paper. As an LTL carrier, Averitt picks up loads from various cities, consolidates them in their regional centers, and then combines common destination loads and ships them to their final destination. With the paper system, 3.5% of shipments were misrouted, driving up costs and causing customer frustration. What Averitt needed was automated entry of receiving and shipping information into its system.

Averitt deployed wireless handheld units to drivers and routing leaders in each of their 92 facilities, so that information is now entered into the system in real time at the docks and integrated directly into their system. Operators receive directions about which dock to take certain shipments, without having to locate paperwork. Errors have been reduced to less than half of one percent, and accurate shipment information is visible in real time. Dockworkers are more productive, receiving movement instructions directly on their handhelds, instead of having to locate a supervisor for instructions. Efficiency has improved, with total freight throughput increasing by 3%, and more goods on fewer trailers makes more money for Averitt.

That's not all. Wireless systems can now track vehicles throughout their entire journey, alerting operators if a vehicle is on an unscheduled stop, trapped in bad weather, or broken down. The real-time information is fed into the master routing and scheduling system, which will alert customers of delays, reroute other trucks, and engage contingency plans. Tracking the location of delivery drivers and service trucks encourages employees to stay on the job. Even more important, it makes dispatch decisions more efficient.

Raco Industries, a wholesale auto parts distributor, installed tracking devices in its delivery vans. According to Richard Gallagher, VP of Raco, the drivers were a little tense at first, but when they saw how efficiently deliveries were being optimized on the fly, the drivers knew the technology had created a competitive advantage.[6]

Penske Logistics, a trucking company, has installed wireless terminals in its 4,000 vehicles as a means of improving the efficiency and visibility of its entire transportation process. In each truck's cab is a unit that allows data input via a keyboard. It also acts as a WLAN, which connects to handheld barcode scanners. The drivers use these devices to scan shipments that are loaded and unloaded. Critical information, such as changes in the delivery order or traffic jams to avoid, are communicated via a satellite network. Other routine information, such as shipment records and logs, are stored in the onboard terminal until the truck enters a Penske dispatch yard. Once a truck enters the yard, the WLAN onboard senses the yard WLAN network and uploads the stored truck information to the Penske central system.

A similar system is used by McLane, a Texas-based grocery delivery business with a fleet of 1,050 trucks and 17 distribution terminals. Before the new system was initiated, communication with the company's 1,000-plus drivers was accomplished by pagers, dispatchers, and a lot of paper. The new wireless system has every driver carrying a handheld unit equipped with a barcode scanner. Drivers use the device to record delivery information and capture electronic signatures. When a shipment is complete, drivers dock their cab-based units in a cradle, connecting them to the onboard communications systems, which uses a dual-mode system of WLAN and satellite link to upload the information to McLane's central system. The system costs approximately $15 million and has an estimated payback of two years, according to Dave Dillon, McLane's Manager of Transportation.[7]

A standout example of a company using wireless technology to solve a huge problem in a very basic business is Cemex, the world leader in cement manufacturing and delivery. The cement business is unique in that once mixed, its only product (cement) is immediately perishable. Cement must be poured within 90 minutes of mixing. If there are delays, the entire batch is lost, and any unused materials must be quickly poured at another client site or poured out on the ground and wasted. Accurately timing deliveries and routing trucks to avoid waste while optimizing the delivery fleet was a huge challenge. Routing the mixer trucks from the mixing plants to customer's building sites, and then routing the trucks to other sites and returning for more materials within the narrow time limits was a logistical nightmare. Cemex installed a computer and GPS receiver in every truck, and then combined positioning information with output information available from various mixing plants and customers' requirements.

Software then calculates where and when each truck should go and how much product it should carry. The system also issues dispatch instructions for redirecting trucks between deliveries. Even with the chaotic traffic conditions in Mexico City and last-minute rescheduling by customers, Cemex has achieved a sustainable increase in the number of orders filled per day and significantly reduced waste.

Wireless has a powerful and growing impact on the shipment and delivery of goods, from groceries and cement to general freight, and soon, even pizza. The Domino's pizza delivery driver will soon arrive at your home with a hot pizza and his wireless-connected PDA. The PDA will have helped him locate your house, allowed you to sign and pay by credit card, and can print out coupons for future discounts. Look for wireless applications in more trucking and delivery applications.

RADIO FREQUENCY IDENTIFICATION DEVICES

Another use of wireless in the logistics chain is the tracking and tracing of assets, both in North America and globally. Companies of all kinds move containers of products and capital equipment around the world on a daily basis, and tracking those assets accurately can save billions of dollars. The asset tracked could be the truck or container itself or higher-priced goods, such as cars or machines that warrant individual tracking. A radio frequency identification device (RFID) is a wireless technology that provides a low-cost way to track shipping containers, larger equipment, or palletized goods. The RFID tags attached to the goods can be read with a handheld reader. The RFID tags are as small as a lighter or as large as a brick and can hold as much as 128K in information, including shipping history, hazardous material information, and a listing of the contents. The tags can also be fitted with GPS technology so they can be tracked from the sky. If not linked to a satellite, the RFID tags are polled at various shipment points, and the data are uploaded through a network to continuously track the shipment. There are also some security benefits because the technology has been modified to work as an electronic lock and history recorder, storing each time the container was opened and for how long.

There are passive RFID tags, which have been around for some time, and now active tags. Passive tags carry less information, have a shorter range, and must be read by more powerful readers. The active tags can have as much as 1 megabyte of memory (roughly equivalent to a novel), but they require internal batteries. The more sturdy ones can last as long as a decade. The passive tags can cost as little as .25 cents each. The price of active tags ranges from $3 to $500 for the most sophisticated tags with memory and power supplies.

The U.S. Army now has an Automatic Identification Technologies branch of its Logistics Automation Division in Europe, which is leading the effort to automate logistics management. The Army is using RFID technology to track shipments and manage the contents of containers. An RFID device is attached to the containers, allowing tracking by a hand-held reader at various shipping stops and at the final destination. Once containers are dropped off at a depot, where hundreds can be stacked in random piles, the devices transmit their codes up to 300 feet to handheld readers. Supply personnel can locate a container and access the inventory through a handheld without having to open and empty each container. This process could have provided a big savings during the 1991 Gulf War, when 25,000 of the 40,000 containers of munitions and supplies shipped to the war zone were opened prematurely simply to determine their contents. In 2001, the Army had tags on more than 200,000 containers of supplies around the globe. A soldier in the field with a handheld reader can point the handheld reader at a container and know exactly what is inside. Another Army project is tagging expensive helicopter parts that are shipped around for disassembly and overhaul. They predict savings equal to the cost of an entire helicopter!

Another company using RFID in logistics is Volkswagen, which places an RFID tag on the windshield of every finished and near-finished automobile rolling off its assembly line at plants in Germany. The cars are parked in massive lots and patrolled by security guards in golf carts equipped with RFID readers and notebook computers, which are wirelessly linked to Volkswagen's central inventory management system. The RFID tags store information on the current condition of the car, remaining accessories to be installed, additional preparations required before shipment, and so on. That information is relayed in real time to the inventory management system for accurate management of vehicles as they are prepared for loading and shipment to dealers worldwide.

To realize all of the benefits of RFID, the technology must be incorporated into supply chain or logistics business process software. A good example of an integrated system is at Associated Foods, a food distributor that delivers roughly $2 million in goods every day to 600 stores in the Rocky Mountain area. Associated's goal in integrating RFID and the software to manage all of the processes was more efficient use of its fleet of 70 tractor rigs and 400 trailers. It also wanted to eliminate waste by keeping tighter control on perishable goods. Now, with RFID tags on every trailer, Associated can locate and track shipping containers anywhere from its 600-acre distribution facility in Salt Lake City, Utah. Data about each trailer, its location, and loading status are transmitted to dispatchers. For perishables, special tags that monitor temperature and fuel in the truck were added. The information is used by dispatchers to schedule a trailer and truck for the next shipment. Tim Van de Merwe, Associated's Logistics Manager, stated that the company has been able to permanently reduce the number of truck drivers from 100 to 62 and the number of leased trailers, which usually handled the unscheduled overflows, from 40 to only five. More than 150 people were previously involved in data entry, walking around the yard, noting trucks, locating special trailers, checking loads, tracking shipments, and scheduling deliveries. With the new software and technology, it takes only two people to manage the system.

RFID technology is just beginning to permeate industry. Applications are numerous and can extend throughout a supply chain and down to smaller and smaller items. RFID fits perfectly into the concept of pervasive embedded machine intelligence (PEMI), where a chip or tag is imbedded into almost everything, allowing communication and visibility of the item almost everywhere. Grocery stores are now using RFID to automatically and instantly update pricing labels, and soon RFID tags will be embedded into the consumable goods containers. Ask yourself: Where can our company use RFID?

IMPLEMENTING THE WLAN

Your IT specialists can address the design issues around incorporating a WLAN into your business. There are quite a few issues and a few points to remember. First, although WLANs provide highly useful communications capabilities, they are like radio transmitters with signals leaking out in all directions. The 802.11b standard has an encryption protocol called wired equivalent privacy (WEP), which is not foolproof. Nonetheless,

WEP should be turned on with the initial rollout. Depending on the nature of the data transmitted, additional security measures should be added, but be careful because they often negatively affect network performance. Consult an expert on WLAN security to address these issues. Second, even though the 802.11b speed is 11 megabytes per second (mbps), which is pretty fast—about the same speed as a corporate hardwired network, a lot of factors determine if you can actually achieve that optimal speed. Because WLANs are shared, not switched, users must share the effective throughput of the particular wireless access point (the antenna). So, for situations where large numbers of users are unpredictably scattered around a site, multiple access points make sense with some balancing technology to keep the traffic jams from occurring and the speed from dropping. The downside is that as you concentrate more access points together, the channel overlap creates contention, degrading overall performance.

When St. Luke's Hospital in Houston, Texas, deployed its new 802.11 system, wireless devices would suddenly lose their signal when elevators containing metal patient beds passed by or when the devices were used in certain corridors near reinforced-concrete stairwells. After an IT investigation to isolate the source, the problem spots were identified. Later, St. Luke's conducted a thorough site survey and has since eliminated all of the problems. The bottom line: Get an expert to help design the system and position the nodes. Insist on thorough testing before final positioning to get it right. Get a guarantee of performance, and test the system carefully before signing off on the project.

REFLECTING ON THE CHAPTER

Applying wireless in warehousing and logistics scenarios produces powerful productivity, process improvement, customer service, and asset management tools. Beginning in the warehouse, then continuing to distribution and delivery, wireless has a potential optimizing role in each segment of the supply chain. No matter what your industry, and regardless of whether you outsource your shipping, manage it in-house, or if shipping is your business, wireless can take it to a new level of productivity, efficiency, and visibility.

ENDNOTES

1. Mike Drummond, "Wireless at Work," *Business* 2.0 (March 6, 2001).

2. Wireless industry statistics, *www.commweb.com/article/ COM20010822S0004.*

3. "HP Connects Mobile Workers and Enterprises with Wireless 'Hot Spots' and New Access Devices," *www.hp.com/hpinfo/ newsroom/press/2002/index.html#jun* (June 24, 2002).

4. Matthew G. Nelson, "Easy-to-Track Deliveries—McKesson's Mobile Devices Get Orders to Customers On Time," *InformationWeek* (September 10, 2001).

5. Mike Krohn, "Ace Beverage: Wireless and Mobile Technologies Reduce Overtime," *Pen Computing Magazine* (November 2001): 53.

6. Matthew G. Nelson and John Rendleman, "Reaching Too Far?" *InformationWeek* online magazine, *www.informationweek.com/story/ IWK2001819S0002* (August 20, 2001).

7. Bob Brewin, "Trucker McLane Rolls Out Dual-Mode Wireless Vehicle System," *Computerworld* (May 1, 2001).

4

Wireless for Customers

Providing your customers with mobile access to information and the ability to transact business and obtain information anyplace or anytime is a powerful catalyst for rapid growth in the coming years. Wireless effectively extends your business hours, integrates your company with the customers' daily operations, and builds an interdependence that increases customer retention. With the correct blending of information and capabilities that match customer interests, a new dimension of customer experience and relationship is possible. This chapter lays out four key guidelines for developing successful customer-facing wireless applications and describes some of the wireless services—and their business benefits—that leading-edge companies are offering to their customers.

Most companies have two kinds of wireless relationships with their customers: informational or a combination of informational and transactional. Wireless informational applications include things like accessing a flight schedule, checking the price on a stock, getting directions, verifying your bank balance, receiving selected news feeds, or sending an e-mail message to schedule a service call on your refrigerator. Informational applications are a great place to start because they are relatively easier, cheaper, and quicker to deploy to customers.

Transactional customer applications are more complex and include things like placing orders for financial securities, concert tickets, and books, and making a rental car reservation. The most sophisticated wireless applications blend both the informational and transactional elements. Companies wishing to extend their business presence to wireless often start small, beginning with the simpler informational applications and then progressing to transactional applications. One thing is for sure: Once the competitive benchmark is set within an industry, competitors follow quickly, and the race for greater wireless functionality, ease of use, and ultimate competitive advantage is off and running.

TRAVEL INDUSTRY

Airlines are setting the pace on wirelessly delivering time-sensitive information to customers. In 2002 up to 70% of the traveling public carried some form of wireless device. United Airlines led the way by providing flight information on the Palm VII in 1999. One year later it followed up with a proactive paging service that notifies passengers when there are changes in their booked flights. Next came a wireless phone application that gives customers access to up-to-date flight and frequent-flier information. United customers register for the wireless service at United's Web site. United's competitors followed fast.

By 2001, Delta Airlines, Alaska Air, American, America West, and Northwest all had similar and even upgraded programs underway. At the close of that same year, 7 of the top 10 airlines had made the delivery of flight status information to wireless devices (text messaging) a standard offering. Today, every airline has some form of wireless messaging. This service wasn't a huge stretch for the carriers, which had recently opened their databases to make departure and gate information accessible over the Web. Pushing information out to cell phones and PDAs was the next logical step as each sought to exceed the others' performance and capabilities. What is important to note is that the competitive benchmark was initially set with informational services, but the transactional performance bar is continually climbing. For example, after American Airlines announced a wireless check-in service—a service high on customer wish lists—several airlines quickly followed.

By 2002, half of the major airlines offered real-time flight status, scheduling, seating availability, key contact information, frequent-flier account information, and the passenger's current itinerary—all over a PDA or mobile phone. The next step is providing wireless flight information and cancellation alerts along with choices of alternative flights and the ability to select one of the alternatives, bypassing lines, delays, and frustration. Notifying passengers of cancelled flights while en route to the airport, plus offering the means to reschedule, provides tremendous value to the passenger. The bar keeps moving. A competitive dogfight is underway as airlines use wireless capabilities to improve customer service and differentiate themselves.

On the ground, Thrifty Car Rental's goal in 2000 was making it simple for customers to book a car, using any manner they chose, including wireless. Today, Thrifty's customers can reserve cars and confirm reservations anytime and from anywhere. Like others in the travel industry, Thrifty had previously offered the reservation service on its Web site, which was used by around 30% of its customers. For less than $100,000, Thrifty extended the Web services to wireless. Commenting on Thrifty's ROI analysis, Brian Carpenter, vice president of marketing for Thrifty, said, "How do you measure the value of giving customers easy access to you?"[1]

The important attribute of "easy to do business with" in the travel industry now includes wireless. As more businesses in this industry offer wireless to customers, the real impact has become visible. The number of travelers using Internet-enabled wireless phones or PDAs in 2002 has surpassed 70%. How are they using their devices? According to a report issued by Forrester in early 2002, around 20% of travelers using mobile phones or PDAs were using them to obtain flight information, select airline seating, and communicate with their hotel and rental car companies. More than 40% were retrieving driving directions and restaurant options, or entertainment information. You can count on continued growth in both the informational and transactional components of customer wireless in the airline, rental car, and hotel industries.[2]

FINANCIAL SERVICES AND BANKING

An early leader is Fidelity Investments, whose first wireless efforts sprung from listening to customer complaints about missing out on investment opportunities because they were away from a wired channel and lacked timely access to market information. Fidelity responded with an informational application called Instant Broker that let active traders monitor their accounts through pagers. In 1999 Fidelity added a two-way capability to allow transactions. In 2001 Fidelity again expanded its wireless services creating Fidelity Anywhere, which allows users to manage funds in their 401K accounts and much more. Fidelity Anywhere had more than 92,000 registered users and was growing by 3,000 users per month in mid-2001. With each iteration of Fidelity's wireless initiatives, the financial services company gained increased customer satisfaction (referrals and retention) and revenue (based on new accounts). Fidelity was convinced that wireless was a competitive differentiator in a highly competitive business, with one-third of Fidelity Anywhere's registered users representing new Fidelity accounts. Fidelity set the pace again by appointing a Chief Wireless Officer, Joseph Ferra, the first on record in the Fortune 1000.

Another financial services success story is e-Trade. The online brokerage was one of the first to extend its Internet-based brokerage services to wireless, offering a Palm application in mid-2000. E-Trade successfully transferred the customer service lessons it learned on the Internet to wireless, notably the value of personalization. In late 2001 Merrill Lynch, Raymond James, and several other major financial services firms followed.

Related to financial services, but less dynamic by nature, is wireless banking. Wireless banking is both informational and transactional because it allows customers to view account balances and past transactions, plus transfer funds and pay bills over their wireless phones. One of the leaders is the Delta Employee Credit Union, which has 30,000 Delta Air Line employees as customer users—and more than half are on the go. The wireless service is an extension of their Internet bank and allows users to check account balances, transfer funds, and make payments anytime, anywhere. The credit union doesn't charge for the service but sees it as a customer retention tool.

Wireless banking has been slow to catch on. According to GartnerG2, a research service of Gartner, Inc.,[3] about 1.2 million U.S. customers used wireless banking in 2002, up from 500,000 in 2001. Although expected to grow quickly, one obvious reason for the slower growth of wireless banking is a relative lack of demand and acceptance by customers. Obtaining bank account information doesn't quite meet the minimal threshold of immediacy that makes it worth using a wireless handheld device to monitor and manage. Unlike stock trading, which is as dynamic as you can get, checking your account balance or transferring funds to cover a large purchase is less immediate. Weak customer response is causing banks to rethink their wireless strategies.

Bank of America put its wireless financial offering on hold in 2002 because of light consumer demand for the service after launching two separate pilot programs. Wells Fargo Bank is proceeding with its wireless offering but is keeping expectations low. In early 2002, Gartner, Inc. reported that only 5% of U.S. banks offered wireless services, but almost one-third of those surveyed planned to offer wireless in 2003.

The good news is that banks that have and continue to invest in wireless do so relatively economically for the most part, by extending their previous investments in online (Web) banking to wireless; however, something more is needed to make wireless banking pay. The Bank of Montreal has also indicated that demand for consumer wireless services is lighter than expected, but it is pressing forward with a new service called "intelligent alerting." The idea behind intelligent alerting is to allow consumers to receive an alert if their account balance falls below a certain level or when their paycheck cleared. Coupled with the alerting feature would be the ability to transfer balances in response to the alerts. In the near future, banking will use interactive voice response coupled with the wireless text alerting to allow customers to pay bills and manage their accounts. Innovations such as intelligent alerting and more advanced voice response systems may be the catalysts for broader consumer adoption of wireless banking.

Another innovation that may help wireless banking is the wireless wallet. The concept of the wireless wallet involves linking a wireless phone account to an authorized credit card. Paying for goods at a wireless wallet-enabled retail terminal involves simply pointing your phone at the checkout terminal and authorizing the charge. The wireless wallet also works with a new type of vending machine. The consumer dials a number posted on a vending machine, their account would be charged, and the item they want to purchase would appear. This is called "m-cash," and it is discussed later in the chapter. In considering the potential acceptance for a wireless wallet concept, the elements of mobility, immediacy of demand, and instant gratification fit the theoretical and practical wireless success model. What may be challenging for this next step in wireless banking is that for now, the benefits do not exceed the security risks, and extra steps are required to complete the transaction from a wireless phone. Wireless banking will progress slowly but steadily as more retail, vending, and bank machines are equipped to interact with the wireless consumer.

LOOKING AHEAD: LOCATION-BASED SERVICES AND ADVERTISING

An important and overdue step for wireless is the incorporation of locater capability into all wireless phones and networks. What's driving this enhancement is a mandate from the Federal Communications Commission (FCC), which requires wireless carriers to install a special e-911 capability that works in conjunction with a global positioning system (GPS) to identify the location of a cell phone within 50 meters, when the phone is turned on. This would allow an emergency crew, for example, to respond to an emergency cell phone call even when the caller didn't know his or her location. Wireless carriers and handset manufacturers had until the end of 2002 to make the enhancements operational, but many have received extensions. Even with the delays and extensions, by the end of 2004 the locater capability will be fully in place in almost every market.

What's the catch? Under current law, tracking a person's location is a breach of privacy, unless it is done with permission. Data collected from cell phone users is protected by FCC rules. These rules require all carriers to keep confidential all user information, unless a customer gives express permission to share it with third parties or, of course, in emergencies.

The business application for wireless advertising is potentially powerful but fraught with questions. In theory, enabling the locater function on your new mobile phone would allow your favorite local restaurant to know when you are in the neighborhood and then push the lunch special information to your mobile phone. Companies at one time were excited by the prospect of pushing advertising to customers strolling through the mall. The expectation was that customers would seek out or pull information into their wireless devices to learn where to buy a smoothie at a discount or uncover a Gap sale. According to early prognosticators, consumers were supposed to flock to services that allowed them to download movie trailers from the Internet as they sat in traffic or ate lunch, but it hasn't happened, at least not yet. Pushing alerts to users is theoretically powerful, but to be practical, such activity must be permission-based and must have some degree of personalization along with timeliness. Pushing a pizza promotion at 9 A.M. is irritating, whereas doing it at 7 P.M. makes sense, but only to those users who have agreed to receiving such promotions.

According to Ovum, Inc., a technology and advertising research firm, the worldwide market for location-based services will grow to $18 billion by 2006; however, the substantial forecast is hinged on proving the commercial viability of location-based advertising services, which largely remain an unproven operational and revenue model. It is not about the technology or the vision of what may be possible. Rather, it is the practical aspects that may limit its success.

Consider what is involved in getting the right advertisement to the right mobile phone customers at the right time. First, a complex web of agreements must be in place among infrastructure providers, carriers, advertisers, and mobile phone users. The coordination required among these parties is extensive and not inexpensive to implement. On top of that, the profiling software is unproven and inexact. More important and fundamentally, as a rule, users don't want to be pushed with advertisements on their mobile phones. Junk e-mail on your desktop is irritating enough!

Proponents of location-based advertising offer a business model in which a consortium of companies gives the phone and wireless service to a customer at no charge, but in exchange for customers agreeing to accept all the push advertising the advertisers think consumers can stomach. This might work for my teenage son, who is always looking for a free deal, but I am skeptical of this model for the mainstream consumer, and especially for the business user. I just don't see most business people selling their mobile phone souls to be abused by advertisers 24/7.

Ultimately, consumers will not want to divulge their location unless they retain control over how and when that information will be used. Consumers may be willing to share their location information and receive information, but only based on their preferences. How much time and effort does it take to fill out the background information on their preferences? Unless you get something for free, who wants to spend 20 minutes or more divulging private information? The bottom line: If you are a retailer and want to engage your customers over wireless, be careful in assessing the real benefits and customer acceptance of location-based advertising. Be prepared to add cost to your project to entice customers to divulge information and provide special access to their personal wireless world.

By contrast, consumers seem to have some interest in pulling information on demand to their wireless phones. In New York, a company named StreetBeam piloted a pull model of location-based wireless advertising. StreetBeam placed infrared beaming devices inside phone kiosks, bus shelters, and on other street structures. A special sign with a flashing light alerts passersby with mobile devices that a beam of information is waiting for them, if they want to download it. The consumer then has the option to point his or her device (infrared = line of sight) toward the flashing sign and coupons, product specials, directions to the retailer, plus a quick link to the retailer's Web site are delivered to the device. Banana Republic is the retailer that sponsored the pilot and, unfortunately for the retailer, the results were underwhelming. Using mobile phones for more than making a call is a learned behavior that has slowly been building steam in the United States. One quick look to Japan gives us a glimpse of what is possible.

The Japanese seem way ahead in making consumer wireless pay, primarily with their DoCoMo brand of colorful handsets, customized ringtones, and short messaging. DoCoMo and Coca-Cola teamed up to allow DoCoMo phones to buy drinks from Coke machines. Members point their cell phones at the vending machines nicknamed Cmo and trigger the release of a drink, which is then charged to the cell phone account. Although the benefits seem minimal, it wouldn't be the first time that an idea was successful just because teenagers thought it was cool. U.S. companies are watching the DoCoMo model closely for continuing consumer penetration, and a few are moving forward with trials in U.S. malls and schools.

OTHER ASPECTS OF LOCATION AND GLOBAL POSITIONING IN WIRELESS

Combining GPS technology with wireless phones, cars, and even watches has some promising consumer benefits. Phones that combine PDA technology as well as GPS locator technology, which inform the user where they are and provide step-by-step directions to where they want to go, are catching on in the United States. These phones support the emergency 911 service requirements for locating an emergency caller and open the door for the true location-based services discussed previously. My personal favorite is providing my teenagers with mobile phones that include GPS trackers that allow me to confirm their exact location; finally, parents get a benefit from technology!

On the subject of tracking family members, there is a huge problem in nursing homes with mature adults who are mobile enough to walk off the grounds but too incapacitated otherwise to take care of themselves. Wearable wireless devices with GPS locating capability are an obvious answer. In addition to the locater feature, wireless personal monitoring units on the market today can detect and transmit blood pressure, heart rate, and temperature. The cost of the technology makes it not for everyone, but the cost should come down quickly. Then again, even the higher cost is attractive compared to having an attendant monitor an elderly person in the home.

Another service which is a near-perfect fit for wireless location-based services is finding restaurants. The familiar *Zagat's Guide to Restaurants* offers information and reviews of more than 20,000 restaurants over the Web and now has extended the data to wireless. Using most cell phones, users can locate a nearby restaurant with only a few keystrokes. The information presented is simple—location, type of food, rating, and price range—but it is exactly what a user needs to make a decision.

TELEMATICS: WIRELESS BEHIND THE WHEEL

In 2001, as many as 1.5 million vehicles were equipped with telematics—the name for automobile-based wireless computing and Web services. Ten times that number will be in place in 2003. An example of an early offering is the OnStar system by General Motors. OnStar automatically contacts your insurance company, the police, or other emergency personnel if your airbags deploy and you don't check in with the system in a certain amount of time. The service can also provide roadside assistance, remote door unlocking, remote diagnostics, and tracking of stolen vehicles. You can even have a concierge buy tickets, direct you to an ATM, or find a hotel. General Motors offers OnStar as an option in its high-end vehicles. Both BMW and Mercedes offer wireless location, emergency, and information services that use GPS as an option on their 2003 model vehicles. Consumers seem to like the service, but it is still early to gauge widespread consumer appeal and marketability.

In addition to the packages offered by the auto manufacturers on new models, add-ons such as wireless theft prevention are available. For less than $1,000, a car owner can have a wireless beacon system installed that rapidly points cops to the car if it is stolen. InterTrak Tracking Services of Frisco, Texas developed the product. Once police identify the stolen car, InterTrak can remotely kill the engine.

Given the early buzz over in-car maps and directions, there is excitement building over extending car-based wireless to include e-mail and Internet access; however, there is a serious intrinsic problem regarding broad use of telematics. There is concern that the complexity of the services will lead to greater driver distraction and more automobile accidents and deaths. By current estimates, driver distractions cause at least 10,000 deaths annually in the United States. A British study released in 2002 showed that talking on a mobile phone while driving is more dangerous than operating a vehicle under the influence of alcohol.

Voice-activated and -operated services appear to mitigate the problem, but no studies as yet have confirmed the benefit. Studies of cell phone–related accidents seem to indicate that hands-free is not much safer than manually operated phones. There is a growing backlash to such in-car services, with many cities, including New York and Miami, prohibiting cell phone use by drivers. Look for similar restrictions on other in-car Internet services, unless workarounds such as advanced voice recognition can be shown to reduce dangers.

WHAT DO CUSTOMERS REALLY WANT?

Consumers can be fickle. On the one hand, they like wireless phones and PDAs that are Web-enabled. The Cahners In-Stat Group predicts that the wireless data market will grow from 170 million subscribers worldwide in 2000 to greater than 1.3 billion by 2004. If the forecast is accurate, more than 1.5 billion handsets, PDAs, and Internet appliances will be deployed. On the other hand, consumers did not readily embrace buying or Web browsing using the early Web-enabled devices. A mid-2001 A.T. Kearney survey of more than 1,600 mobile phone users in the United States and Europe found that only 12% expect to engage in m-commerce transactions—down from 32% the prior year. More telling is that the survey revealed that only 1% of users had actually made any purchases with their phones in the prior year. The challenge is that customers expect a positive, efficient, and seamless wireless commerce experience, and the 2G networks, devices, and applications did not deliver. Proponents believe that faster 3G networks will improve the customer experience to a level that will rapidly drive consumer adoption. Would they say anything else?

An interesting element of the survey was mobile users' interest in using their Web-enabled phones as a substitute for cash, also known as m-cash. In the survey, 44% of users indicated interest in making small purchases by dialing a phone number that corresponds to a product delivery station like a vending machine, gas pump, or parking meter; however, although interest appeared high, only 2% of people surveyed in late 2001 have actually used m-cash and not many more did so in 2002. This is largely because the infrastructure is not yet in place to allow pervasive m-cash transactions. Why not? The business case for m-cash infrastructure is weak because the cost to retrofit vending machines and parking toll stations is significant compared to the incremental increase in revenue. It is too early to tell if m-cash will be the killer application or if, like wireless Web commerce, the expectations will exceed the early utilization.

There is potentially great value in consumer wireless, but it is tough to achieve just the right mix of technology and practical usage. The U.S. Postal Service discontinued its wireless program, after investing millions of dollars in it. The wireless application allowed consumers to track packages, locate the nearest post office, and find a zip code. Sounds nice, but these services sound more applicable to a Web offering than wireless because they lack the inherent elements of immediacy or time-sensitive information that is necessary to attract users. The services were not aligned with what consumers need or want when mobile. Customer-facing wireless programs at Fidelity, Thrifty Car Rental, and *Zagat's Restaurant Guide* were successful because they deployed simple applications that hit the sweet spot of providing timely information to mobile consumers and delivering an acceptable user experience.

BUILDING THE CUSTOMER-FACING WIRELESS APPLICATION: HOW TO PROCEED

For most companies, creating a cost-effective and relevant customer-facing wireless application will be a serious challenge. The key to a successful customer-facing wireless application is not necessarily the whiz-bang of the technology or functionality (more on this in later chapters); rather it is the usability from the consumer's perspective. Although there are no hard and fast rules, there are a number of suggestions you should consider in the process of developing a customer-facing wireless application.

○ *Commit to including customers in every step of the development process.* Doing so takes more effort and planning, but it is the single most important component of the development process and the primary driver for success. The unofficial reason for most consumer-facing wireless failures is not that the application doesn't work properly, but that the consumers don't like it or won't use it. Gather together some customers and share your wireless ideas and assumptions

with them. The sweet spot to look for is using wireless to more efficiently mimic an existing nonwireless process. You may be surprised
to learn from consumers that you have defined less or more than
they want or need. Having customers at the design table injects a
continuous dose of reality into the development process. Customers
will help keep your development team focused on what they want.

○ *From the initial customer discussion meeting, pick a few friendly customers
(ask for a volunteer) to work with you as partners in the development and
testing.* These customers will be at the design table from the early concept meetings and help you conduct the necessary iterative field trials.
Field trials are where real customers use the wireless application with
the sole purpose of determining what works and what doesn't. Start
the field trials as early as possible. When you have the bare basics of
the application completed, expose it to your field trial customer, and
let them pound on it. Then adjust and tweak the application as you
build the next revision. Repeat the field trial process until the field
trial customer gives the application a thumbs up.

○ *Create a collaborative and open environment where your IT development
leaders are in the room with your customers as they describe their feelings
about the application.* Encourage customers to criticize the application: Tell them it's all right to say it's junk! You would much rather
hear it during the development cycle than read it in *The Wall Street
Journal.* Remember, information presented via wireless looks different than on PCs. Display screens are smaller, memory is limited,
and keyboarding is miniaturized or not available. It takes a lot of
trial and error to get it right. Don't stop the field trials until the customer users are happy—I mean really happy. Do not serve the wireless wine before its time, even if it takes longer than you planned.

○ *Now you are ready for the test market process. Be selective. Do not launch
your new application to all of your customers at one time.* Instead, roll
it out to a sample set of customers. Also, select a representative
sample number of your field salespeople to support the launch. Make
it a manageable group of 50 or fewer people. In the test market
process, you are learning the reactions of a broader base of customers, plus getting a sense from your field salespeople of the
application's usability. Establish the threshold of user adoption that

equates to success. Measure the results carefully and be realistic. Call the customers who rejected the application and ask them why they did so. This step is actually more important than rejoicing over the customers who accepted the application. Ask them what is missing in the application that would make them use it. Find out precisely and consider making the suggested changes. With the customers who adopted, explore why they chose to adopt: Do they find real value in the application and are they willing to deploy it throughout their company, or was it just an interesting test? The bottom line here is don't confuse an early adopter interest as broad interest. You are now finally ready to launch!

DEFINING THE ROI

Defining return on investment (ROI) for any technology investment is essential. This topic is discussed in detail in Chapter 6. As a basic primer, ROI must consider the total cost of development, which includes the cost of acquiring and developing the software, the personnel costs to conduct the field trials and test market, the cost of training for support operations, and the promotional materials and advertising to launch the product. These costs should be amortized over a period no longer than 12 months based on the rapid changes in wireless technology.

The ROI analysis is often calculated based on the projected costs and benefits of the project and is relied on in making the go/no-go decision. Then it is usually shelved. Consider having quarterly ROI reviews, in which ROI analysis is verified or adjusted at various points in the development process, because it is critical to know the good or bad news earlier rather than later. Keeping the ROI current will also help in calculating the total cost of ownership and ongoing operational costs.

BENCHMARK THE COMPETITION

Six months after United Airlines launched its wireless flight status service, Delta announced a wireless check-in service. The competitive bar for wireless functionality will continually move up for every industry. Check out your competitors every month or so, observing and assessing the features and functionality they offer. Don't fall in love with your current application. Be ready to build more features and functionality into the next version of your application, but only after the user demand is confirmed.

The preceding guidelines are basic, but they will help prevent you and your company from overspending and/or alienating customers with a fancy but incomplete application—or even worse, a useless application. Developing a successful customer-facing wireless application is not easy. The sooner you get started with a basic application, the easier it will be to deploy when the competition surprises you with theirs.

REFLECTING ON THE CHAPTER

The rewards of successfully extending the business value proposition to mobile users are new customer acquisition and increased customer retention; however, deploying the customer-facing wireless application is challenging and carries a good deal of risk. The benefits of extending your brand and accessibility to customers through wireless can be a game-changing move, as it was for Fidelity, or it can be an initial loser, as it was for the U.S. Postal Service. You may be able to wait on deploying customer-facing wireless and concentrate on your internal operations. If you are an airline or car rental company, wireless for your customers is now just one more price of admission because of competitive offerings.

The pace of customer-facing wireless development is definitely picking up; new applications are announced every month. The march to expand customer-facing wireless continues in the banking, vending, advertising, and transportation sectors. Which industries will be next? One day, your customers will ask for it or let you know that your competitor just provided it to them. It is an eventuality that every customer will use wireless for some purpose at some time in the future.

ENDNOTES

1. Alan Radding, "Leading the Way," *Computerworld* (September/ October 2001): 16.
2. "What Wireless Travelers Want," *MBusiness* (February 2002): 10.
3. Kristy Bassuener, "Wireless Banking to Double in 2002," *WirelessWeek.com* (March 4, 2002).

5

Wireless in Health Care, Government, and Education

Some of the most prolific and productive deployment of wireless computing is occurring in health care, government, and education.

WIRELESS IN HEALTH CARE

In health care, more than 800 million doctor-patient consultations take place annually, and 98% of them are handwritten, collected, and stored on paper. Much of the paper is generated at hospitals, fast-paced centers where health care workers move around regularly, making them among the most mobile professionals in the workforce. In the midst of all the paper and the hoped-for healing is a troubling number of errors. According to a 2000 report issued by the Institute of Medicine,[1] medical errors are the eighth leading cause of death in America, with some significant portion of the errors flowing from mistakes in recording or interpreting written diagnosis and prescriptive information. The cure for eliminating some of the errors may be wireless.

Hospitals and physicians are adopting a variety of wireless applications, ranging from checking in patients via wireless electronic tablets to accessing medical records, writing electronic prescriptions, and wirelessly tracking critical equipment that is moved from floor to floor. The results are promising.

A focal point for error reduction is the drug prescriptive process. Paper-based processes are giving way to electronic prescription writing. PDAs can provide doctors with instant access to patient records and reference materials on complex drug interactions and produce legible prescriptions to transmit to pharmacies. In physicians' offices, electronic prescription transmission is one of the fastest-growing uses of wireless. Instead of scrawling what often looks like chicken scratch on a pad of paper, many doctors now use a stylus or minikeyboard to write a prescription. They synchronize their handheld devices to a server, which faxes and e-mails the medication order to the pharmacy. According to a study by Gartner, Inc., in June 2000, only 5% of physicians wrote prescriptions electronically, but a projection by the W.R. Hambrecht group forecasted that usage would quickly increase to more than 20% by 2004.[2]

Blue Cross & Blue Shield of Rhode Island inaugurated an electronic prescription program with 42 primary care physicians in the state. The doctors were given Palm Pilots and a free subscription to an online drug database for six months. In addition to helping reduce errors in translating prescriptions, the software has a huge database to flag dangerous drug interactions. In another case, Blue Cross & Blue Shield of Rochester, New York, offered physicians $100 toward the purchase of a Palm Pilot to be used for electronic prescription writing, and 400 doctors accepted. Supporting the program are drug makers Eli Lily and Bristol-Myers Squibb, which offer doctors Palm Pilots with prescription-writing software in exchange for information about their practices and an agreement to accept certain advertising on the Palm Pilot.

Three powerful drivers are behind the trend in electronic prescription writing: error reduction, drug interaction avoidance, and automation of the paper process. Industry studies show that electronic prescribing reduces medication errors by 55%—more than half! Drug interaction databases are published annually in paper form, but electronic databases are updated daily. When a popular drug is withdrawn from the market, electronic prescription databases alert the doctors immediately.

Health maintenance organizations (HMOs) are excited about the technology because they could influence the database by referencing lower-priced generic drugs. Premera, a large medical insurance company in the Northwest, offered a free PDA program downloadable from its Web site. The program allows doctors to know how Premera classifies the drug they are considering prescribing, and it will suggest lower-cost alternatives. The program prevents the doctor from prescribing a drug that is not approved on the insurance plan, avoiding delays at the pharmacy and costly calls back to the doctor's office for an alternative. The program also has a powerful drug interaction database, allowing doctors to check for drug interactions on up to 30 drugs simultaneously. Premera believes the program will lower its member out-of-pocket costs and pay for itself very quickly.

Harris Interactive reported results of a study of doctors' wireless priorities as follows:[3]

53%	Immediate access to lab results and records
43%	Ability to generate notes (for medical records) more easily
26%	Ability to record billing coding information more easily
14%	Immediate access to clinical content during consults
12%	Immediate access to trusted protocols and guidelines
7%	Immediate access to pharmaceutical formularies
7%	Ability to generate an authorized referral based on insurance plans

As you can see, the top three requests relate to being able to access and interact with hospital records, and many hospitals are considering new electronic medical records software (EMR) that will allow doctors to wirelessly access data. Automating doctor access to patient records helps simplify administrative processes, which in turn means more productive and less-stressed health care professionals. By adopting wireless technology, they can also make more money. Surprisingly, doctors lose about $60,000 annually as a result of missed or lost billings, according to Synergy Medical Informatics, a benefits management firm. Physicians must record one or more diagnostic codes after diagnosing and treating each patient. Getting the correct codes and all applicable codes is a key to accurate billing and reimbursement. Even the more conservative U.S. Department of Labor confirms that before adopting an electronic alternative, the average physician loses between $35,000 to $50,000 because of miscoding.

Physicians working at Dedham Medical Associates, a private practice near Boston with 75 physicians representing 12 different medical and surgical specialties, would enter the various codes on a multipart paper form. The forms were turned over to the billing group, where the data were entered again into a computer billing process. Errors in entry, transcriptions, and missed billing opportunities were costing plenty. Now physicians enter the complex codes on their Compaq Ipaq pocket PCs, then wirelessly transmit them via a 802.11b wireless network to the central billing system. With special software designed to walk physicians through complex coding rules, the wireless system has improved accuracy and streamlined the process, eliminating the double entry system. Errors in miscoding and data entry have decreased by more than 50%.

Doctors and hospitals are moving toward mobile computing simultaneously; eventually practice management systems and electronic medical records will be integrated. Mobile computing for doctors is a win for the hospitals, drug companies, physicians, and their patients because it works to improve care and increase accuracy and efficiency in recordkeeping.

On the hospital side, Moses Cone Health System in Greensboro, North Carolina, installed a system in 2001 to speed processing and reduce errors in updating and retrieving patient records and transmitting and fulfilling prescriptions. Doctors use handhelds equipped with infrared ports to access patient records, including up-to-the-minute lab results, and to download treatment plans from synchronization sites located around the hospital. Doctors can regularly update their PDAs by synchronizing them with the hospital system throughout their shifts. Moses Cone cut their errors by half.

Moses Cone CIO John Jenkins said that the system cost around $250,000, all of which the hospital expects to recover over the next few years based on the significant reduction in errors and time savings for doctors.[4] The system eliminates the need for doctors to sit down at workstations to review patient records and to print and fill out required forms. Doctors say that the system saves them anywhere from 30 minutes to an hour per day. Pharmacists also save time because prescriptions arrive more quickly via the system, instead of on paper. Drug interactions can be quickly referenced and incorporated into patient instructions. Doctors who practice in the region but who are not on the hospital staff are also allowed access to the system. Dr. Matt Martin, a trauma surgeon who practices at a private clinic but performs surgery at Moses Cone, commented, "Technology helps people when it's easy to use, intuitive, and you can pick it up without a learning curve."[5] When Dr. Martin arrives at the hospital, he synchronizes his Palm Pilot and uploads information about all of his patients, their medical records, and current lab results. "It saves me time logging on to the system and scrolling through screens."

Some hospitals, such as Beth Israel Deaconess Medical Center in Boston, are extending wireless access to patients. The hospital serves 60,000 patients per year, providing some patient wards with wireless Internet access via laptops so patients can surf the Web. In addition, the hospital extended wireless to its emergency room check-in process. Before, patients in pain would have to drag themselves to the desk and provide check-in information to a clerk. Today at Beth Israel, the clerks come to the patient's bed and gather the information on a wireless tablet. Patient information is updated on a new electronic dashboard in the emergency department command center for doctors and nurses to review. Replacing the old whiteboard and scribbles, the electronic dashboard displays key patient data in real time, along with lab values and other information. As doctors examine patients, they input information into their wireless laptops, which updates the hospital system and dashboard.

Soon, voice recognition systems will make data entry even easier. At the Albert Florens Storm Eye Institute in South Carolina, physicians piloted the use of voice recognition software that lets them speak the patient information (diagnosis and vital signs) into their handhelds, which transcribes the information to the screen. Once the examination is completed, the physician then reviews and accepts the record, which is sent to a central database for inclusion in the master patient record. As early as 2001, IBM teamed up with Mdoffices.com, offering a service that lets doctors record patient notes using spoken commands on a pocket PC device. The doctors' voice recording is collected and stored as a standard audio .wav file, which is then transcribed using an IBM speech recognition product. The notes are edited by a medical transcription service and sent via e-mails or faxes to patients' pharmacies or labs. The service is priced at $2.50 per transcription. Doctors pay around $3 per page for transcription from recorders today.

Doctors aren't the only health care providers to use wireless. Traveling nurses who provide in-home services are also benefiting from the technology. The Visiting Nurses Association (VNA) in Santa Ana, California, is leading the way in using wireless to support home health care. A shortage of nurses prompted the VNA to look for a way to reduce paperwork. According to its analysis, every hour of health care delivered to a patient generates 30 to 45 minutes of paperwork to document and bill for the services rendered. An average mobile nurse visits between five and eight patients per day, and busy nurses would need to take files home in order to stay current.

The VNA decided to write its own application for the Palm operating system. The first application was for documenting each patient visit and for time and attendance reporting. As VNA staffers became more adept at using the forms-based application development tools, they developed applications for scheduling, ordering supplies, infection control reporting, IV drip calculators, and more. All forms feature simple lists, and each day nurses upload that day's patient list, schedule, time and attendance records, patient demographics, and physicians' orders. When traveling, the nurses' Palm Pilots can be connected via wireless or they can be synchronized at night when the nurses return home. The synchronization updates the medical records with that day's patient treatment information.

The VNA also incorporated a drug interaction database that holds information about more than 460,000 drugs. Patients average 10 medications each and often are receiving drugs from several practitioners, who may not be aware of each other. More than 100 nurses use the system, which translates to a 50% reduction in paperwork or three hours per day! Do you think there is a future for wireless and home health care? Definitely.

Wireless computing in health care extends to hospital and medical practice suppliers as well. Owens and Minor, a supplier of operating room supplies, has tapped wireless to automate the reordering process for surgical products. Owens and Minor equipped the operating rooms of dozens of its hospital customers with a predetermined level of surgical supplies, all of which are labeled with a barcode. A staff member uses a wireless handheld to scan items as they are used during surgery. Data are then transmitted through a wireless collection point and on to Owens & Minor's central system. The system adjusts the inventory, bills the hospital, and sets up the reorder.

Watch for even more growth and greater adoption of wireless computing in all areas of health care. Technologies or practices that reduce errors and improve patient outcomes begin as good ideas, then once proven become the minimum standard of care. Fear of litigation is a motivating factor. If it was proven that using a handheld with a drug interaction database linked to patient records eliminated errors, and the doctor and/or hospital chose not to utilize the technology, they would have to explain why they did so when being sued over a wrongful death resulting from drug interaction origins. Besides the liability aspect, productivity improvements go a long way in health care, where staff shortages are prevalent. Wireless computing utilization will deepen and become more commonplace in health care.

WIRELESS IN GOVERNMENT

Federal, state, and local governments are deploying wireless in a big way. Would it surprise you that in Washington D.C., the Feds have a secure channel for their BlackBerrys so Generals and Congressional representatives can stay connected in a secure manner? Surprisingly, governments, which are often slow to adopt new technology or ideas, are beginning to embrace wireless and mobile computing, and they are among the technology's frontrunners.

If you dig a little deeper, it started at the top with the 1996 Government Paperwork Elimination Act (GPEA). This act stated that government forms have to be online and fully automated by the end of 2003. To ensure the deadline was met, in October 1999, President Clinton put a booster on the GPEA with a Presidential directive stating that the 500 most commonly used federal forms must be online by the end of 2000, which was substantially met. President Clinton also signed the Electronic Signatures in Global and National Commerce Act—E-Sign or the E-Signature Act. Taken together, the GPEA and E-Sign have pushed the U.S. federal government to the Web, and it is not a stretch at all to move from accessing information over the Web to accessing it via wireless.

The trends extend to the military as well. The U.S. Army has created an Automatic Identification Technologies branch in its Logistics Automation Division in Europe. In 2001 the Army had Radio Frequency Indentification Device (RFID) tags on over 200,000 supply containers around the globe. In addition to the RFIDs, the Army is using wireless to aid in the recruiting process. Wireless is used to streamline the process for tracking candidate inquiries—triplicate paper forms are replaced with electronic wireless forms that can be completed by recruiters on the fly during the critical graduation period. The system forwards candidate inquiries to field recruiters within 24 hours instead of two weeks. Now potential recruits begin their journey to "be all you can be" wirelessly!

Local governments are catching the wireless bug as well. The Transportation Department of the city of Fairfax, Virginia, equipped 12 transit buses with GPS devices linked to an automated vehicle location system that transmits real-time location and scheduling information to passengers' mobile devices. The information is viewable by Internet phones, PDAs, and on bus stop monitors. The system is designed to delight riders, boosting ridership and freeing up support employees who used to frequently answer bus-scheduling questions. The system also allows managers to monitor drivers' whereabouts and on-time performance.

In the city of Richmond, British Columbia, flood waters threaten the city from time to time making quick reaction to water level alerts and pumping data critical. By replacing their manual system with wireless, city crews can wirelessly access a database that contains up-to-the-minute information on water pumps, water levels, temperature sensors, and sewage levels via 180 field sensors that blanket the city. Preset alarm levels trigger alerts that are pushed out to the crew's wireless handhelds to allow immediate response.

In Miami, Florida's Dade County Building Department, field inspectors have wireless handsets that allow them to submit their inspection reports to the city's central server for immediate posting. Building contractors can access county inspection results as soon as they are posted, allowing them to take action to correct problems or move occupancy forward. The system has eliminated a two-day wait between inspection and posting and, more important, it has eliminated the need for permit clerks to rekey the report information into the construction permit databases.

In San Mateo County, California, Fire Chief Richard Price wanted firefighters to have access to information that could help them save lives and approach fires more efficiently. Price championed providing the firefighters with RIM pagers and Compaq Ipaqs to allow them instant access to the county's fire dispatch database. Price is using the firefighter handhelds as both input tools and management tools, keeping track of firefighters across various assignments and in large fires, and pushing them information and instructions. The fire trucks were originally equipped with rugged laptops that cost more than $10,000 to acquire and install. The handhelds cost one-fourth of that, plus they can be removed quicker for mobile use.

Wireless is also becoming a powerful tool for state and local law enforcement agencies, including the Illinois State Police (ISP). With more than 3,000 officers and civilian workers, the ISP is responsible for providing law enforcement throughout the state, often supporting small and remote communities with backup policing. Yet its legacy radio communications systems were more than two decades old and coverage was limited to major metropolitan areas. The ISP's goals in deploying a new wireless mobile computing and communications system were straightforward:

seamless statewide coverage, immediate access to information in key law
enforcement databases, and ease of use for officers in the field. The ISP
chose the cellular digital packet data (CDPD) wireless protocol for data
transmission. This option uses only the unused bandwidth on the analog
channels of wireless phone networks offered by AT&T and Verizon. These
CDPD systems run on existing cellular towers that offer 19.2kbps data
speeds; however, they provide a low-cost and reliable network. The ISP
saved millions of dollars by avoiding having to maintain its own network.
Security was incorporated via multiple levels of encryption. Patrol cars were
outfitted with rugged laptops and high-powered wireless modems. Most of
the hardware and software was off the shelf, including the barcode reader
for quickly scanning drivers' licenses and a small thermal jet printer for
receiving low-resolution black-and-white photos of suspects. The units also
have GPS capability, allowing the central dispatch office to track the location
of patrol cars. The ISP successfully upgraded the coverage and functionality
of its communications network using existing technology at reasonable prices.

WIRELESS IN EDUCATION

Schools are havens of information, schedules, and communications with a
highly mobile population—perfect for wireless applications. That's why
schools from the elementary to university level are pioneering wireless in
various forms. Best stated by Eric Peterson, assistant headmaster at Forsyth
Country Day School, "We see wireless as a transforming technology for
our students and for schools in general."[6] Every member of the third-grade
class at Forsyth has a Palm Pilot. The young students have quickly learned
how to work the devices and use them to retrieve homework assignments
and take tests. The seeds of the next wireless generation are being sown.

Universities caught on early to the benefits of a wireless campus.
Installing wireless local area networks (WLANs) throughout a large campus
keeps the thousands of undergraduate and graduate students connected to
the pulse of their academic environment and prepares them for a wireless
future.

In spring 2001, the University of South Dakota became the first university in the United States to require handheld computers for first-year
medical and law students. The university provided 1,300 Palm handhelds
to these students. The students' Palm Pilots are loaded with financial cal-

culators, reference materials, literature, coursework, and organizers. The university built a high-speed infrared LAN system to connect the handhelds to the university computer system. Access points are available throughout the campus, in classrooms, hallways, the library, and labs. Students can locate faculty members, retrieve e-mail, and look up e-mail addresses. The university estimated the initial cost for the overall system at about $40 per student, excluding the cost of the Palm Pilots.

Wireless networks provide an excellent way to improve campus communications, with minimal construction and aesthetic impact to the facilities, many of which have significant historical value. Providing network access throughout the dozens of historical buildings on the 116-acre Boston College campus was the challenge for Henry Perry, Director of Network Services, and network engineer Brian David. What convinced them to deploy wireless was the punitive cost and negative aesthetics of ripping out ceilings and stringing cable through historically important and tough-to-access buildings. Faculty and students can now use laptops to access the school's 802.11b wireless network from more than 350 wireless access points throughout the campus. Students can register for classes, view curricula, and use e-mail from wherever they may be on campus. WLAN technology is fast becoming the standard for any first-tier university.

Wireless is also helping the school systems operate more efficiently. In the Boston public school system, approximately 63,000 students are truant at some point. Boston public uses wireless to support the truancy officers who track the students as they wander the city during school hours. Boston public's truancy officers spot likely truant students and ask them for identification. Before the new system, officers would then pull a phone book–sized document from their cars and leaf through thousands of pages attempting to locate the student's record. Later, paper forms would be filled out and sent in for input to update the student's record. Today the phone books and paper forms are gone. Truancy officers now carry Java-enabled phones that quickly access an online repository of student and legal records, such as warrants. Truancy officers can access and update student records all during a student encounter. The new phones have a two-way fast-connect feature as well, so officers can link to each other quickly. In the future, the school district plans to add barcode readers to scan student IDs, speeding information input. Truancy officers can check more students per day and keep the records updated with less overall effort.

REFLECTING ON THE CHAPTER

Hospitals, doctors, pharmacies, and home health care services are rapidly increasing their use of wireless in the daily patient care processes. Lives may be saved and costs will go down. Governmental agencies from the U.S. Army to state and local police and wastewater treatment departments are using wireless to improve the flow of information and stay connected to the pulse of the community. Dozens of universities and even K–12 campuses are deploying wireless to connect their students to campus information. These institutions are embracing wireless computing technology to solve real problems and reach new levels of productivity, responsiveness, and efficiency—and we will be the beneficiaries.

ENDNOTES

1. Institute of Medicine Report, from "Medical Errors: The Scope of the Problem: An Epidemic of Errors" as reported online at *www.ahcpr.gov/qual/errback.htm*, provided by the Agency for Healthcare Research and Quality, Rockville, MD. Information also taken from "Handhelds and Healthcare" by Rick Broida and Jessica Hardwick, *Handheld Computing* (July/August 2000): 35.
2. John Edwards, "Mobile Medicine Starts Now," *MBusiness* (January 2001).
3. *Ibid.*
4. Howard Baldwin, "2001 Mobile Master Awards for Enterprise Deployments," *MBusiness* (January 2002).
5. *Ibid.*
6. "PDAs Rule the Schools" *Pen Computing* (November 2001): 60.

6

Mapping Your Company's Wireless Strategy

Every CEO of the First Practice companies discussed in this book had to answer the question now facing you: Should we deploy wireless now or later? Each of the First Practice companies answered "Now!" But what process did they use to move forward with a strategy and a plan? This chapter offers guidance and perspective to business executives who are considering a wireless initiative.

Consultants often are skeptical of simple steps or checklists, but many businesspeople have learned from experience that simpler is often better. I would recommend approaching your wireless initiative in four sequential phases (see Exhibit 6.1). Each phase addresses a key issue, with several critical questions that must be answered and tasks that should be accomplished before advancing to the next phase. The four phases and associated key issues are as follows:

1. *Evaluation.* What should our company do with wireless?
2. *Planning.* How will we accomplish the goal?
3. *Development.* How will we acquire, develop, and integrate the required applications and supporting infrastructure?
4. *Implementation.* How will we deploy, support, and measure the wireless program?

PHASE 1: EVALUATION—
WHAT SHOULD OUR COMPANY DO WITH WIRELESS?

The evaluation phase is by far the most important in this process. In this phase, the details regarding the audience, processes, goals, cost, and the

Exhibit 6.1 Extending the Enterprise to Wireless Computing

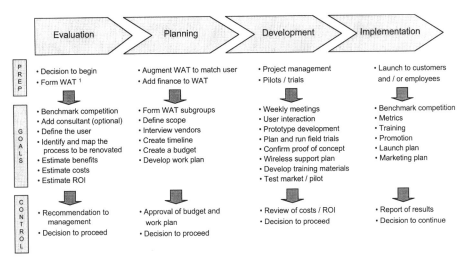

¹ WAT— Wireless Action Team

estimated ROI are defined. Developing the answers to the four key evaluation questions outlined as follows will help you frame the fundamentals of your company's wireless computing plan. Upon completing the evaluation phase, it should be possible to decide whether to go forward or wait to begin the wireless initiative. To properly evaluate the opportunity for wireless, you need a cross-functional team of employees—the Wireless Action Team.

Forming the Wireless Action Team

There are several rules to follow in forming the cross-functional Wireless Action Team. The team should not exceed seven people (don't ask why), not including the executive champion, and should never have more IT people than non-IT people. The ideal makeup of the team is one person from each of the following disciplines:

- ○ Sales
- ○ Operations
- ○ Marketing
- ○ Accounting

- ○ IT legacy (enterprise) operations
- ○ IT Web operations
- ○ IT database operations

Select the facilitator from this group, or use an outside consultant facilitator, and then begin the first phase. Time and expense can be saved by shortcutting the Wireless Action Team process, but you will pay for it many times over later.

In Phase 1, you must answer four evaluation questions in order to lay the foundation for the practicality of the wireless application and the ROI modeling necessary to avoid disappointment. This is laborious but necessary work. The questions are:

1. Who is the wireless user?
2. What business processes will the wireless application address and what is the improvement or value added to the process?
3. What will it cost?
4. What is the return on investment to the company?

Who Is the Wireless User?

The first question to consider is: "Who is the target audience for the wireless rollout—employees, customers, or suppliers?" For competitive reasons, some companies in our First Practice 50 (the case study companies included in this book) chose to create a customer-facing application first; however, most have not. A 2001 study by IDC Research revealed that of the companies initiating wireless programs, a whopping 82% said their target audience was employees. Only 44% were targeting customers and 19% were targeting suppliers. The numbers quoted add up to more than 100% because they reflect an overlap where the surveyed companies were targeting more than one audience with their initiative. Still, the employee-first scenario is clear among the First Practice 50, with 70% rolling out their first wireless applications to their own workers. Several unique issues exist around deployment of wireless to employees, and different issues exist when deploying wireless to customers. Choose wisely and for the right reasons.

What Business Processes Will the Wireless Application Address and What Is the Improvement or Value Added to the Process?

Wireless doesn't add value unless it improves a process, such as field sales support, customer service, order entry, fulfillment, or a back-office function. Generally, the higher the value added, the higher the complexity and associated costs. Exhibit 6.2 illustrates the relationship between complexity, value, and cost. Most employees and customers resist learning new technologies or processes unless they see a clear benefit to doing so. This is especially true for customers who have a choice to continue doing business with you, even if it's not via wireless. After you have identified a few key processes that may benefit from wireless, take the time and map them out carefully, step by step. Carefully defining the business processes and value added will help the organization make the best decision regarding investing in wireless or deferring the wireless program.

To keep the discussion practical, remember that the real value in wireless computing is realized when the person served by the process is not able to access a fixed computer for some critical period. Another key factor is that the process that they seek to access should be time sensitive or enhanced by immediacy. Third, significant value is possible when wireless

Exhibit 6.2 Wireless Computing Value Curve

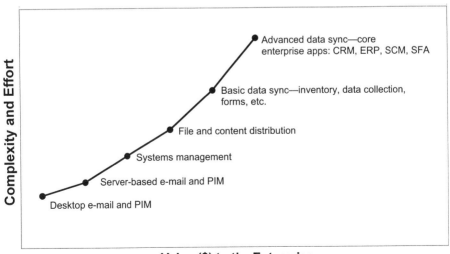

Source: Synchrologic

can eliminate a redundancy in a process, such as rekeying ordering information. To add value, a wireless application must address at least one of these three criteria. Developing a wireless application that addresses all three is even better. Remember these three criteria for wireless applicability:

1. Access

2. Immediacy

3. Efficiency

I learned the access lesson when my former company took our customer-facing wireless ordering application to a Ford Motor Company plant. The application was originally designed for contractors who didn't have access to a computer while at remote job sites. We were so excited about how well the application worked that we wanted to show it to Ford, one of our key industrial customers. The purchasing execs loved it. Then we took it down to the real users—the maintenance workers whose job it is to keep the plant running. They thought it was a handy tool; however, the lead tech questioned me: "Why would I need this to access my inventory and ordering info when there is a Ford ERP system kiosk every 100 feet throughout the plant?" He was right of course. Why use wireless when you have easy access to a faster fixed computer? We thanked them and headed home. We had forgotten about the three tests of practicality for wireless, especially the one about access.

Information with high immediacy is defined as information that is both time sensitive and dynamic. Stock quotes and airline schedule updates are two prime examples. The guiding principle would be as follows:

> The benefit from obtaining the information sooner rather than later must be greater than the cost of the wireless technology and the time spent obtaining the information through alternative channels (like picking up the phone).

Can your salesforce and/or customers benefit from having certain information always available to them, from anywhere? If so, what is the information and how will it be used? Consider if there are immediacy benefits to accessing real-time inventory data, dynamic pricing, order status, or sales records. Apply the benefit versus cost formula to help define the answer.

Efficiency is the third and final key criterion for wireless applicability. Wireless is not the answer if it takes longer than doing something by phone, fax, or fixed computer. The exception would be if a redundant process step is eliminated or the overall process time is compressed. Most employee wireless solutions are aimed at process improvements, which yield measurable productivity gains. For example, Mitchell International, an insurance services support company, has hundreds of claims adjusters working in the field. The claims adjusters use laptops with wireless connections to pull down vehicle and driver information while on a claim site, saving trips back to their office PCs and reducing overall customer claim processing time. Mitchell stated that the wireless technology has increased its claims adjusters' productivity by 20% to 30%.

Transactional is more difficult, higher risk, and if successful, very meaningful to the business. Here is a checklist of typical business functions to consider:

- Customer interaction
  - Lead tracking
  - Quotations
  - Inventory availability
  - Review of order history
  - Return material authorizations
  - Order entry
  - Order confirmations
  - Order status updates
  - Advanced shipping notice
- Sales and sales management
  - Lead generation
  - Customer opportunity alerts
  - Call reporting
  - Contact management
  - Advertising specials
  - Promotions

- ○ Logistics management
  - ○ Warehouse/inventory management
  - ○ Kitting (prepacking of like materials)
  - ○ Receiving
  - ○ Counting/picking/shipping
  - ○ Integrated shipment tracking (link to UPS/FedEx)
  - ○ Delivery truck routing and tracking
- ○ Accounting and finance
  - ○ Proof of delivery (e-signatures)
  - ○ Credit card approval
  - ○ Credit card processing
- ○ Customer service
  - ○ Instant access to always updated customer records
  - ○ Filing claims and refunds
  - ○ Requesting credits

Wireless deployments of every one of these business applications listed have been pioneered by the First Practice 50.

When you have selected the functional area and business process where you wish to apply wireless, consider that the existing processes were built around historical practices and technologies. Many companies have vestiges of old processes embedded into the current processes, and these could be streamlined or removed. Would it surprise you that some of the processes your employees use are not meaningful? For a reengineering project at my former company, we interviewed various users to determine what databases and scripts they were running to do their jobs. We wanted to make that sure all of the affected processes were included in the reengineering effort. Several employees gave us a list of programs and steps they were using each day. We checked, and although some of the programs had been obsolete for years, the employees we spoke to had continued to run the application. We then asked them why they were running each one, and surprisingly, several answered that they didn't really know why! Their only response was that the person training them for the job had told them to do it, and they just continued doing it that way for years and years. Some of their output was not being used at all by anyone.

Remember this formula: NT + OP = EOP. Translated, this formula means new technology (NT) plus old processes (OP) equals expensive old processes (EOP)! Incorporating wireless with a focus on business process is a great opportunity to refine, rethink, or even reengineer current processes. There may even be opportunities to completely remove work steps while retraining the workers to use the new wireless tools. Look into all aspects and processes of the business for possible uses of wireless.

For Phase 1, question 2, you must fully define—step by step—the value added and process changes necessary to your wireless application. You may choose to stage features over time, deploying simple functions first, and then increasing complexity over time. It isn't practical to simply transfer all of your Internet or network applications to a cell phone or PDA. A serious amount of refining must take place. What are the key components that must be sent, received, or accessed by wireless users? It is useful to label these components as informational or transactional to help identify how difficult it will be to incorporate them into your wireless strategy. Informational is easier, less risky, and potentially less meaningful.

Although starting small is important, it is equally important to consider the optimal end state and include that thinking in planning beyond the short-term application focus. Defining the long-term feature roadmap is important in order to avoid unintentional design constraints (more about this later).

What Will It Cost?

In surveying some of the First Practice 50's wireless projects, their costs range as widely as the depth and scale of their applications. For example, UPS spent more than $100 million while the City of Richmond, British Columbia, spent $100,000. To estimate what an enterprise wireless program will cost, you must consider several variables. To better understand the overall scope of what is required to extend the enterprise system to wireless, review Chapter 7, Extending the Enterprise to Wireless. After doing so, the following cost checklist will make even more sense:

- Direct software and development costs
  - Mobile devices
  - Mobile application servers
  - Middleware
  - IT programmer time to develop/customize wireless applications
  - IT programmer time to interface the wireless applications to the business systems
  - Upgrading uptime on key servers and communications links
  - Security software and integration
  - Consultants
- Indirect costs
  - Management time to plan and implement
  - Staff time to implement (affected operational departments)
  - Staff time to fully test security measures
  - Cost to develop training and rollout materials
  - Additional training of help desk personnel
  - Additional help desk personnel
- Cost of ongoing operations
  - Lost productivity during changeover to new processes
  - Monthly telecommunications charges
  - Mobile device management costs
  - Cost of issuing devices (security passwords, setup, etc.)
  - Repair/replacement of handheld devices

Each company's wireless deployment represents a unique combination of these cost elements, and the total cost will depend on the scope and scale of the planned application. Therefore, until the desired application is defined in detail, there is no one number. After completing and documenting the Phase 1 evaluation questions 1 and 2, and after addressing the previous checklist, consider hiring an outside consultant to help estimate the costs. Keep the engagement very narrow, and ask for one or more alternatives for each component of costs, including dividing the project into steps of complexity and corresponding steps in costs. The larger the overall scope of the application, the greater assistance that may be required to estimate the costs.

What Is the Return on Investment to the Company?

Defining the tangible and intangible ROI for an enterprisewide wireless deployment is formulaic once questions 1 through 3 have been answered and the process and value-added maps have been defined. The tangible ROI comprises the hard dollars that are achieved in one of three ways:

1. Improving productivity (doing more with the same resources)
2. Lowering operating costs (costing less to produce the same output)
3. Increasing sales

As an example, Gartner Inc., a technology research firm, estimates that the total cost of ownership of an enterprise PDA system is between $1,800 to $4,300 per user per year. This would include the cost of hardware, connectivity, maintenance, and support costs, but doesn't include the initial development and testing costs. Assuming field personnel cost $100 each per hour, it would take 18 to 40 hours of realizable productivity savings per year, or the equivalent net benefit in sales increases, just to break even with the cost of deploying the wireless application. This is only 30 to 60 minutes per week per employee over a year and is probably achievable based on one or more of the following productivity benefits from wireless:

○ Quicker access to up-to-date customer records (fewer phone calls)
○ Less time filling out and tracking paper forms
○ Fewer trips back to the office to file documents and access records
○ Less time and paperwork to place orders
○ Fewer errors and error correction steps resulting from fewer data entry steps
○ Fewer errors because of improved adherence to structured processes
○ Increased productivity through better use of downtime

Wireless can often affect all three of the performance variables (i.e., productivity, cost reduction, and sales increases) because the benefits of process improvements often cascade throughout an organization. The Sears repair team anecdote described in Chapter 2 offers a great First Practice example of the cascading benefits effect. Sears HomeCentral employs 13,000 field service representatives, who are able to make an additional 9,000 calls per day, thanks to their wireless system. The productivity increases are realized by allowing the mobile workers to constantly input data and stay connected with dispatch, eliminating delays between repair calls and other downtime. Service technicians get replacement calls and routing information, which leads to more time in the field making customer service calls and generating revenue.

Another benefit has been the productivity improvement in ordering parts. Before having their wireless system, Sears service technicians placed 1.4 million calls per year to order parts. The calls were estimated to cost between $2 and $4 each. With the new wireless system, which bypasses the call center, call volume and call center headcount has been reduced by 80%. A positive impact was made on cash flow as well because service technicians began billing customers on the spot, processing transactions via credit card instead of mailing a bill later.

What began as a dispatch process improvement led to improved field technician productivity, which allowed more sales calls, resulting in higher revenue. Also, downsizing the call center effectively and permanently lowered costs. Finally, cash flow improvements were made possible by quicker billing. Sears kept the cascading effect going by reengineering the entire inventory and ordering system for parts—all keyed from the field service technician's wireless terminal.

Lowering operating costs can come in several forms, including reduced inventory, reduced staffing, reduced overtime, or improved cash flow (lowering working capital). Office Depot deployed wireless technology that electronically captured customer signatures on delivery and then transmitted the completed shipment information and signature to the company's central system for immediate billing. By speeding billing, Office Depot improved its cash flow by several days, which in turn avoided millions of dollars in interest on working capital. By carefully mapping out the processes and value added, the benefits should naturally bubble to the surface, allowing a reliable ROI estimate.

The most exciting consequence of improved productivity and a reduction in business process cost is greater sales. Getting to make one more customer call per day per field representative to get one more order goes right to the bottom line. Many of the First Practice 50 were able to estimate increases in sales attributable to their wireless deployments. These companies include Ace Beverage, Jim Hudson Lexus/Saab, Celanese Chemicals, Producers Insurance, and of course Sears.

As an ROI benchmark of tangible or direct savings, a study conducted in 2002 by Iain Gillot Research surveyed 35 major companies using wireless computing applications for a variety of purposes, including e-mail, personal productivity, salesforce automation, workforce automation, dispatch and routing, and customer relationship management. The companies were asked to provide their best estimate of direct (tangible) savings for their application. Direct benefits reported included increases in service calls of 32%, reduced customer wait times by as much as 80%, increases in sales of 10% to 20%, and service responsiveness improvements of 7%. Most of the payback times were cited to be less than six months, although some were more than 30 months.

Each of the First Practice 50 illustrates a tangible benefit in at least one of the three performance areas and can be used as benchmarks to potential savings. But what about the intangible benefits that are more difficult to measure? These intangible elements are more closely related to brand building and differentiation, both of which must ultimately lead to increased sales in order to have value. Many intangible benefits result in improved customer satisfaction. Although customer satisfaction is not the easiest variable to measure, happy customers usually buy again and again. Customers are happy when they get a prompt response to their requests and their orders are handled right the first time. The intangible categories include the following:

○ Increased customer retention (existing customers leave less often)

○ Increased customer mind share (deeper account penetration = customer market share)

○ Competitive differentiator (a reason to choose one over another = new customers choose us more often)

A great example of both increased retention and competitive differentiation is Fidelity Investments. In October 1998, Fidelity launched its first wireless service called Instant Broker. Fidelity customers were missing out on investment opportunities because they were away from a wired channel and didn't have access to market information. Instant Broker let active traders monitor their accounts through pagers. In 1999 Fidelity added two-way capability to allow transactions and expanded the information feeds. Today, Fidelity has expanded its wireless service even further, creating Fidelity Anywhere, which allows users to manage their 401K accounts and much more. In its first year of operation (2001), Fidelity Anywhere had more than 92,000 registered users and was growing by 3,000 users per month.

Fidelity's Chief Wireless Officer, Joseph Ferra, addressed the question of ROI this way: "Our overall objective is better customer service, which is difficult to place a value on. However, about one-third of Fidelity Anywhere's registered users represent new Fidelity accounts." He continued: "Once a customer subscribes, they stay. They really like the convenience and control."[1]

With each new wave of wireless services, Fidelity has seen increased customer satisfaction (referrals and retention) and revenue (based on new accounts). Fidelity's choice was not so much whether to offer wireless to customers—because competitive pressure had established the expectation—but how many wireless features to offer and when to offer them. Fidelity decided to lead its market segment instead of following, using wireless as a competitive differentiator. Wireless is a reason to choose Fidelity over the other guys.

The Fidelity example illustrates a great point about ROI: Think long-term program, not project. Like Fidelity, you should probably roll out wireless features and services in waves, starting first with basic functions and then enhancing the capabilities based on your experience, market requirements, and user adoption. Plan on a three- to five-year program and prove the ROI at each step.

Keep in mind from the outset that saying "no" or "maybe later" to wireless is not necessarily a mistake. The First Practice 50 have experienced success, but not every wireless project is a resounding success. A lot of time and money can be spent on wireless without a favorable outcome.

Consider Carlson Hospitality Worldwide, the parent company of Radisson Hotels. In its first attempt at wireless, Carlson chose to develop a customer-facing application. It spent hundreds of thousands of dollars on piloting high-speed wireless access in its key hotels, but the project was stopped after six months because only 3% of guests used the service, which had an installed cost of about $500 per room. The ROI simply wasn't there.

In its second attempt, Carlson concentrated on deploying wireless to its employees, notably hotel and key department managers. The second deployment was cost effective and operationally meaningful because occupancy levels increased and overtime was reduced. Carlson learned a valuable lesson: To avoid failure and wasting a lot of money, before the pilot or even before hiring a consultant, think through all aspects of wireless as it will be applied to your specific business scenarios. In short, answer the four key questions in Phase 1.

Congratulations, you have just completed Phase 1 and the four key questions, including the key ROI analysis. Once Phase 1 is completed, the Wireless Action Team should submit its findings and estimates to the executive champion for review and discussion. A go/no-go decision should be sought from senior management. Let's assume your team gets a notice to proceed, and let's move on to Phase 2: Planning.

PHASE 2: PLANNING—
HOW WILL WE ACCOMPLISH THE GOAL?

With the four questions of the evaluation phase answered, you are now ready to outline the workplan or project plan for your company's wireless initiative. Many companies have formal project planning and management processes that are routinely used, and these should be used for their wireless project as well. The following are some ideas to include in the project planning process and the ultimate workplan.

Leave in place the Wireless Action Team throughout the planning phase, but augment it with people who have hands-on experience in the area targeted for application of wireless. If you plan to bring wireless to your warehousing operations, include a warehouse manager on the team. You might also want to include the appropriate IT application development supervisor. The augmented team can now approach the next step of resource planning.

For the non-IT members of the team, have them review Chapter 7, Extending the Enterprise System to Wireless. The chapter defines the four distinct segments of work, which blend together to form a successful wireless application. Providing them with this orientation will help keep the nontechies from losing perspective.

Resource Planning

I recommend defining and managing the project based on the four layers of work required as explored in Chapter 7, Extending the Enterprise to Wireless. Those layers are enterprisewide applications and databases, mobile middleware and synchronization, telecommunications networks, and wireless devices and applications. Forming two- to three-person subgroups for specific planning on each layer works well. For example, if your wireless application is customer facing, add sales and customer service personnel to a team focused on the wireless device and application layer. These people work as a subteam to ensure the usability of the application. In addition, IT legacy personnel would be assigned to the enterprisewide application and database layer. You may need a consultant to help your most qualified IT people with the middleware layer and the security issues. Don't get me wrong: The teams must be tightly linked and coordinate closely to achieve a common goal and coordinated application; however, segmenting and focusing on excellence in each layer will yield a better application.

By now the team has a rough idea of the extra personnel required, including consultants, who need to be a subset of the team. A reminder: If the wireless application is customer facing, you must be sure to add a customer, or minimally, get some customer input at some time in the process. If you skip this step, be prepared to pay more and to take longer in rolling out your initiative.

Vendors

You will need software and telecommunications vendors to help you with some portion of the initiative. Early in the process you need them for cost and time estimates. Once your staff has a grasp of the details of the work, consider what can be done inhouse. You may be pleasantly surprised by how much your inhouse staff can accomplish if given the time and training. My best vendor contracts were for specific deliverables of a project, together with training of our staff programmers. This allowed us to make future changes and upgrades on our own.

Most companies have some vendor selection process or criteria used to qualify and then compare vendors and their proposals. Consider scheduling a vendor day, where the top qualifying vendors give back-to-back presentations of their product and experience. During the presentations, ask for specific examples of meeting timelines, budgets, and creating special value for the client. Ask for detailed customer references and confirm each of them. Ask the references if they have received any benefits (such as reduced prices or free service) to provide references for the vendor. Also, during the presentation, keep careful notes of the claims made and carefully confirm them with the references provided. When you arrive at the price, be aggressive in negotiating, especially for software. If you pay more than 60% of the original asking price, you paid too much.

The golden rule in working with vendors is, the more you know what you want, the easier it is to get it at a reasonable cost and on time. Keep your scope of work narrow and clearly defined. Resist scope creep because every vendor wants to sell you everything. Don't trade vague promises of seamless interoperability for proven performance. If you have an applicable vendor relationship already established, it often makes sense to include that vendor on the Wireless Action Team, at least in one or more subgroups.

Timeline

Estimating the overall timeline from start to final deployment of wireless is downright scary. Here are two suggestions. First, after each of the four layer groups is formed and defines its work steps, have them estimate their individual timelines. All of the timelines should be integrated into a Gantt chart system. (Microsoft Project is pretty easy to use and inexpensive.) Then align the timelines to incorporate the sequence of handoffs. Add plenty of time for coordination between groups, at least an extra week or two. Add an extra two months or more for field trials and beta testing. It often takes as much time to test applications as it does to develop them.

Second, if the application is customer facing, add at least three more months for a narrow test market period. You can't afford to roll out a second-rate application to all of your customers. It is more important to get the application right than to have it deployed exactly on time, but managing a timeline with precision is a necessary part of corporate discipline. The timeline helps create the sense of urgency needed to keep the subgroups moving forward toward their common goal.

Budget

Revisit your initial cost estimates from Phase 1. Have all members of the Wireless Action Team review the cost estimates and sign off on their areas. If it is a customer-facing application, add additional programming hours for final upgrades and corrections at the conclusion of the test-marketing period. Spend some time discussing worst-case scenarios, and include a contingency amount of 10% to 20% for surprises. Deploying a successful application that costs twice the amount of the approved budget will hurt your credibility and cause a recalculation of the ROI.

Like all project plans, yours will undoubtedly change, probably sooner than you thought possible, but the power in any plan is more than good practice and cost management; it allows for a better end result and allows the managing of organizational expectations. Also, remember that the more novel and challenging a project, the higher the risk of budget and timeline variance. A detailed workplan will help limit your risk. Last, although senior management may have approved the project when you completed Phase 1, take the time to develop a solid project plan and timeline before you actually start work.

Skipping forward, your project plan is now completed, most of the blanks are filled, and the teams are formed and ready. You obtain buy-in from the executive champion and then give the word to begin development.

PHASE 3: DEVELOPMENT—
HOW WILL WE ACQUIRE, DEVELOP, AND
INTEGRATE THE REQUIRED APPLICATIONS AND
SUPPORTING INFRASTRUCTURE?

This phase in the project is where the IT development work gets done, but it creates the most anxiety for the non-IT personnel (like upper management). Although time passes and money is spent, it appears that little is being accomplished. The programmers have their heads down and are diligently toiling, vendors are scurrying about smiling and saying optimistic things, and your boss passes you in the hallway and gives you a thumbs-up and smiles, exclaiming "wireless—oh yeah." As the head of the wireless project, you go back to your office and quietly fret. After all, you have virtually nothing against which to measure your progress. Right? Not quite.

As the wireless project manager, you should have progressed sequentially through Phases 1 and 2 which would have required each working group to define its work steps and timeline. Schedule weekly meetings to get project updates and schedule adjustments from each working group. Ask each group the same direct question: "Are you on the timeline and on budget?"

If you are not the project manager or team leader, make sure whoever is has weekly meetings and updates the project timeline. Request demonstrations of any components of the software that are completed. Get samples of the possible devices and pass them around for testing. Test everything, early and often. Then test it again.

The development phase is also an ideal time to plan the field trials and test-market exercises to take place during the implementation phase. Field trials are part of the development process where users (either customers or employees) test drive various components of the application and provide feedback on usability and possible improvements. This is also an ideal time to plan the full implementation, including training, documentation, and support materials that would be used in the actual rollout. Once the wireless application is ready and field trials have confirmed its functionality and usability, you are ready for a real pilot program (for employee use) or a test-marketing process if it is customer facing.

If it is a customer-facing application, ask the sales and marketing team members to recruit customers for the field trials (beta tests). This is where you approach customers to use the product in the normal course of business. The test-market process allows you to confirm user acceptance that goes beyond the easy-to-please early adopters to the more conservative users who require a clear cost benefit. Use the test market results to forecast overall user demand, cost to deploy, and price sensitivity. Recompute the ROI using the test market data.

In the field trials process, you should also begin preparing a mobile device management (MDM) plan that defines how to support, issue, and repair devices if required. Include the procedures for security protocols and password management. Prepare FAQs for employees or customers. Support staffers can test the MDM plan during the field trials and test market.

Finally, use this time to plan, train, and equip your IT and customer help desks to support the new application. If it is a customer-facing application, responding immediately to customers' help requests will be imperative. How will the help desk handle hardware or service provider problems? Include these issues in your project critical path rollout plans. Most important, you can use the help desk as part of the information-gathering process on the field trials.

Ultimately, you must know how users feel about the application both as a tool and with specific functionality. Get the specifics. Challenge the initial assumptions with a monthly reality check session. Do users like the idea of the application or the application itself (an important difference)? Take the time to list the negatives and any bad news from field trials or tests.

If all goes as planned, the application fine tuning, integration details, field trials, test marketing, rollout, and support plans will converge near your estimated launch date.

You are now ready to launch.

PHASE 4: IMPLEMENTATION—
HOW WILL WE DEPLOY, SUPPORT, AND MEASURE
THE WIRELESS PROGRAMS?

The test-market data have come back with good results. The team initiates a few tweaks to the application or documentation, and the wireless initiative is ready to launch. If it was a customer-facing application, you have completed a test market representing a sample of your customers. Assuming that customer feedback was positive, you are ready to implement wireless in your organization and/or to customers. If the evaluation, planning, and development phases were properly executed, this phase will be very satisfying.

Most problems that occur will be around training (How do I do something?) or communication (Who do I call?). To ensure success, request status reports from the help desk and field personnel every day for the first few weeks.

You may also uncover the need for an unprecedented level of network and system uptime. When thousands of devices are accessing your organization anytime (possibly all the time) and from anywhere, uptime becomes everything. Imagine requiring all sales personnel to access the company's pricing files before quoting a customer, then the wireless server goes down, effectively shutting down the entire salesforce. Not good. That is why it is absolutely critical to build in a program of metrics, which include uptime measurement, as well as other performance metrics. Last, don't forget the security issues. Check and recheck their operational effectiveness and include a security review in your ongoing monthly performance metrics analysis and reporting.

Depending on the scope and scale of each company's wireless initiative, a dedicated wireless program manager may need to be employed to carry the initiative forward. Fidelity Investments, a true pioneer in the technology, has a Chief Wireless Officer.

You may think it is time to celebrate, and it probably is, but make it quick because at this point you should circle back to the evaluation phase and have the team conduct a postlaunch review of the entire process, including development and operating costs and ROI assumptions. The bottom line is the ROI impact. Did you meet or exceed the projection? To thine own self be true!

Congratulations! You have successfully launched your wireless initiative, but there is no time to rest. A few months after the successful launch is a good time to begin planning the next wave of your wireless offerings, beginning again with the evaluation phase. Recall the Fidelity example in which each year for three years Fidelity gradually ramped up its wireless capabilities. Most companies will find a staged approach to feature growth makes the most sense.

REFLECTING ON THE CHAPTER

The four key questions in the Phase 1 evaluation are the fundamental starting points to mapping your company's wireless strategy. Choosing to deploy to employees or customers is a defining decision with many implications. Then begins the hard work of carefully documenting the business process where wireless will be utilized, along with evaluating the real value-add or process improvements to be achieved. Finally, examining the various aspects of ROI will establish the objectives and expectations for your wireless deployment.

Phases 2, 3, and 4 provide the frame of reference for planning, developing, and implementing your wireless initiative with discipline and purpose.

At the risk of sounding like a broken record, trust the process of forming a cross-functional Wireless Action Team, then allow it to take the time required to carefully evaluate and plan the process. If you do, the chances of winning increase exponentially.

The next chapter will help put the scope of complexity of a wireless project in context of the enterprise business system and should be helpful to non-IT personnel. That chapter is followed by the top 10 lessons learned by some of our First Practice 50 in developing and deploying their wireless applications.

ENDNOTE

1. Alan Radding, "Leading the Way on Wireless," *ComputerworldROI* (September 2001).

7

Extending the Enterprise to Wireless

As an executive of a company embarking on a wireless project, you need to know about the basic technical issues involved in extending the enterprise's functionality to wireless users. You will be at the table when the company's IT team, department heads, and the consultants assemble and begin to discuss the project details (e.g., how the wireless deployment will be accomplished, how much it will cost, and how long it will take). Other key questions will include how much in-house and outsourced work is needed and what portion of the enterprise system must be modified or augmented to enable the wireless deployment.

To prepare for the discussion, you should become familiar with a few basics, such as:

○ A typical wireless enterprise system architecture and functionality
○ More details on networks, devices, and their operating systems
○ Security
○ The mobile device management plan
○ Support considerations

When you complete this chapter you should have a good idea of the various components of a wireless enterprise system and be able to participate in the planning discussions.

THE ENTERPRISE WIRELESS ARCHITECTURE

An enterprise wireless program has four interlocking layers, and each layer has four key factors to consider. Most of the work, the challenges, and the problems can be categorized into one or more of these four layers. The key issues in each layer are:

Exhibit 7.1 Wireless Enterprise Architecture

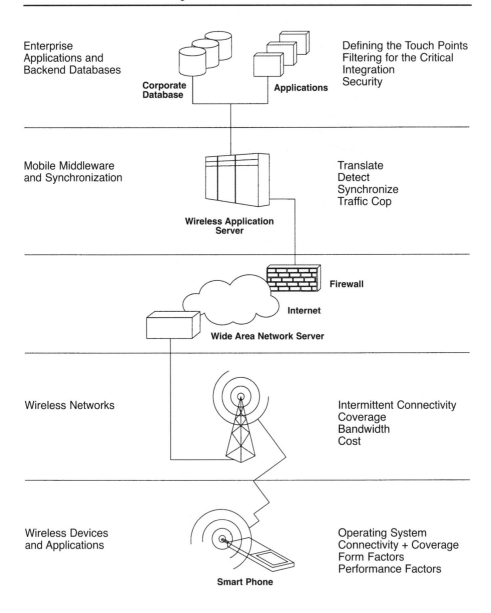

Enterprise Applications and Backend Databases

Corporate Database

Applications

Defining the Touch Points
Filtering for the Critical
Integration
Security

Mobile Middleware and Synchronization

Wireless Application Server

Translate
Detect
Synchronize
Traffic Cop

Firewall

Internet

Wide Area Network Server

Wireless Networks

Intermittent Connectivity
Coverage
Bandwidth
Cost

Wireless Devices and Applications

Smart Phone

Operating System
Connectivity + Coverage
Form Factors
Performance Factors

ENTERPRISE APPLICATIONS AND BACKEND DATABASES

Every functional component of an integrated wireless application maintains touch points or integration points with the enterprise system—either a corporate database or a backend process. For example, when a customer checks inventory by accessing the corporate inventory database, the accessing of the database to satisfy the request from the application represents an integration point. Similarly, a customer who enters an order into a business system through a wireless device would require several integration points. If the Wireless Action Team has defined the planned wireless application in great detail, it is straightforward to define the necessary integration points to the primary (enterprise) business systems. In defining the integration points, it becomes clear where the enterprise system will have to be modified and the scope of the role of supporting software.

If you have invested seriously in extending your enterprise information to the Web, you are probably a step ahead, at least with some informational applications. Developers have learned that integrating Web-enabled applications is relatively easier than non–Web-enabled apps, primarily because the Web applications are already built on a three-tier Internet architecture (i.e., presentation layer, business logic layer, data layer). The wireless application would add one additional layer at the device and extend certain components of the presentation layer. Large business system applications often integrate these three layers, making it more difficult to isolate and extend the presentation layer for wireless unless it has already been done for the Web.

Even with an extensive Web presence however, be prepared to reassess the user experience when the Web site is accessed via wireless. There may be some subtle differences between hardwired Web and wireless Web in the areas of log-on time, latency (slowness) in page changes, and distortion or clipping in graphics.

The process of extending a Web application to wireless is called *transcoding*, and it is relatively inexpensive to outsource or it can be readily learned in-house. Make sure to use an open standards approach like Java or XML to give your systems maximum flexibility in the future. For more than basic information that involves updating databases or interactive applications (like ordering goods), however, additional software and some customized programming are required. For real-time transactional wireless integrations your Web site may be more of a hindrance than a help, so don't assume that the Web equals wireless.

MOBILE MIDDLEWARE AND SYNCHRONIZATION

Enabling the incorporation and synchronization of real-time data from wireless into the enterprise business systems is a formidable challenge. A genre of software tools appropriately called middleware is available to help bridge the gaps between the enterprise system and the nuances of wireless networks and devices. Middleware is the trade term for one or more software tools that translate and manage information as it moves from the corporate database to the wireless device and returns. Middleware performs four key functions:

1. Detects the protocol differences of inbound device queries and adjusts to fit the standardized enterprise criteria

2. Synchronizes the wireless application data input to the business system operations

3. Acts as a traffic cop to hold and forward information to and from wireless devices when the network connection is broken or the business system is unavailable

4. Translates the enterprise applications to a form that wireless devices can utilize

Individual middleware point solutions for each function can be purchased separately, but beware of interoperability issues. Multifunctional middleware is available as well, but beware of overstatements by vendors touting complete solutions. You hear the message: Beware of software vendors! The IT group on your Wireless Action Team should take its time and do its homework in evaluating and selecting the providers of middleware because it is critical to the success of the wireless initiative. Let's review each of the four key functions of middleware.

Getting Wireless Data into ERP

Receiving queries from multiple types of wireless devices that may change over time requires detection and adaptation middleware that adjusts to the software formats for each particular type of wireless device attempting to access the application. Translating corporate information and application data for display on mobile devices requires translation software. This type of middleware adjusts for different sizes in wireless device screens, different wireless markup languages, and different communication protocol nuances. It sounds pretty complicated, but it may soon become a lot easier. Large companies including Microsoft and IBM, are offering wireless application gateway (WAG) software, which is billed as an out-of-the-box solution for accommodating the various wireless protocols, networks, and integration hooks for most premier systems and devices. The WAG standards are sponsored by 50 interested companies, including Oracle, Microsoft, and IBM. The concept of using a wireless gateway over the Web is aptly identified as a wireless portal.

If you choose the wireless portal concept in lieu of a middleware software package, the integration and synchronization requirements between your company's business system and the wireless portal middleware are still required; however, the benefit is you only have to do it once (you hope) as the portal provider will keep up with the latest new devices and formats. Because of the rapid changes in capabilities and pricing, wireless portals are a viable option. Carefully review and compare the wireless portal outsourcing option against an in-house solution. The key factor is the anticipated variety of wireless devices, both now and in the future, that will be accessing and operating with your application. The larger the number of different types of devices, the more the wireless portal concept makes sense.

Synchronization and Session Management

A large corporation gave out thousands of Palm Pilots to its employees as gifts. Everyone was pleased, except the IT department. The next week the IT department was crushed with requests to integrate and synchronize the new devices with the corporate technology infrastructure. Welcome to one of the most challenging issues in utilizing wireless devices for your employees—synchronization with your corporate system. The issue becomes even more complex when your employees have multiple devices, as they invariably may, ranging from Palm Pilots and Ipaqs to Web-enabled cell phones, smart phones, BlackBerry handhelds, and any other number of devices yet to hit the market.

There are two styles of synchronization: sometimes synchronized and always synchronized. Which one you choose will make a huge difference in the level of technical complexity and cost. In defining the synchronization criteria for the wireless applications, you should determine your company's tolerance level for wireless applications that do not work while out of the coverage area.

If you choose sometimes synchronized, the emphasis shifts to the device and the application running on the device. You still need the middleware for synchronization, but the wireless application can run while the user is out of coverage range. Later, the user connects to the corporate system and updates databases in the corporate system and in the device. Although this approach is easier and more reliable from the perspective of the wireless user, it still requires sophisticated synchronization software to update all of the appropriate databases, execute the appropriate applications on the central system, and then update the device. The optimal middleware can help with store-and-forward features, message queuing, and accessing multiple databases simultaneously. For salespeople and managers, e-mail is usually the first application to be sometimes synchronized. After that, personal productivity information, such as contacts, to-do lists, and calendars are added. It can also work for orders and other forms-based applications.

Domino's Pizza offers an example of a sometimes synchronized wireless application. Domino's quality control auditors visit franchises across the country, conducting audits and surveys. Before the auditors go out to the field each day, they connect and synchronize with the Domino's corporate systems by loading their Palm Pilots with surveys and specific information about each franchise they will visit. Their Palm application works in the offline or stand-alone configuration; they don't need a wireless connection to conduct and record the results of an audit. The handheld unit only requires communication with the corporate system to upload a completed report and to download a schedule or other information. When an audit is completed, the local store manager can sign off on the visiting auditor's findings right on the auditor's Palm device. Using a wireless modem on the Palm Pilot, the auditor then sends the completed report back to the Domino corporate system and downloads the next audit assignment. Once the download is complete, auditors can do their jobs without having to be connected and synchronized with the corporate system. A sometimes synchronized capability works well for Domino's auditors.

If you choose an always synchronized configuration, you need an always-on connection almost all of the time. Allowing wireless employees to interact with corporate data and applications to enter an order, change a delivery date, or accomplish other work requires more complex software tools. Occasional interruptions in the connection, slowness in the interaction between the device and corporate system, and a dozen other factors require a middleware buffer between the wireless devices and the business system that is specifically designed to support the more complex always synchronized operation.

In tandem with synchronization is the concept of session management. Business system user sessions that are initiated through a wireless device, then interrupted by loss of signal, then restored again, can be kept open with the proper middleware to manage the wireless session. Accordingly, always synchronized is a real challenge. Although the telephony networks are improving every year, it is still a little risky to depend on a real-time, always-on connection to the corporate business system; however, as the stability and coverage of networks improve, it may become easier to deploy an always synchronized (real-time) application.

This overview of middleware is an oversimplification, but it should at least convey a sense of the functionality required to connect and control real-time wireless integration. Depending on the range and performance of the applications extended to wireless devices, the need for synchronization varies widely but is an essential element of transactional wireless.

Even with steadily improving middleware, the factor most affecting the continuity of the wireless connection is the network. Better networks mean less dependence on middleware for session management and buffering, and less need for stand-alone device applications.

WIRELESS NETWORKS

The reliability and performance of the communication link between your corporate system and your wireless users are critical. The link is also the one element that is completely out of your control. Accordingly, recognizing the weaknesses in the networks and planning around them will increase the chance of success.

One of the biggest problems you will face in sorting through the network issues is session reliability. Session reliability relates to the loss of data or transactions caused by interrupted communication sessions. If you listen to the wireless network carriers, reliability is improving, but would you bet a customer or key shipment on it?

An alternative to improve reliability for mission-critical functions is to blend two or more technologies together into a dual mode or hybrid network that provides redundancy and ultimately near 100% reliability. This approach is obviously more expensive, but it is necessary to achieve uninterrupted and reliable wireless operation. Most hybrid systems include terrestrial or land-based communications provided by a carrier together with a private satellite backup.

The real promise of wireless computing is hinged on faster data networks. Network performance defined in terms of speed and throughput capacity is a key factor in the usability and operation of your enterprise wireless application. Better network performance equals faster execution of wireless applications and a timelier user experience. Speed directly affects the user experience and may heavily influence the ultimate adoption (and ROI), especially in customer-facing applications. Why take 20 minutes to accomplish a task via wireless if you can do it faster by alternative means?

The upgrading of the U.S. telecommunications infrastructure will have an important effect on business. The next generation of higher-speed transmission protocols allows for up to a tenfold increase in voice bandwidth—the bread and butter of the carriers—but the big deal about higher-speed wireless networks is what they mean to wireless computing.

Network speed is the bottom line. Consider the time it takes to download a 1,500 kilobyte PDF file on a wireless PDA. With the second-generation (2G) systems we've had for the last few years, it takes roughly 22 minutes. With the newer 2.5G speeds, it will take only five minutes. Using the third-generation (3G) speeds, now available in most of the larger cities, it only takes 30 seconds. Almost always, higher speeds translate to greater user satisfaction. Higher speeds and application responsiveness also translate to greater functionality, such as wireless video and other streaming media.

Another improvement in the new networks is the always-on functionality, which eliminates login delays and lets providers push information to users at any time. For continuous e-mail, permission-based marketing, and flash updates to work, the always-on feature is required. Check carefully, though, because not all upgraded networks have this push capability. Last, the new data transmission protocols allow network operators to track data usage by subscriber, forever altering the telecommunications data billing paradigm from per-minute to per-megabyte charges. You will primarily pay for the data you are moving, not for the time.

High-Speed Networks

AT&T, Verizon, and Sprint, plus a few others, are committed to completing and eventually broadening the infrastructure for 2.5G and 3G. They have spent billions over several years to make the deployments in major cities and must achieve some level of ROI before adding smaller cities. Accordingly, I would not recommend basing your wireless strategy around a working nationwide 3G network until mid-2004 or later. If you need a nationwide network, what are the alternatives?

Two solutions are well underway and in place in a growing number of smaller cities. They are named 2.5G and 3Glite, and they effectively split the difference between 2G and 3G speeds; both are a lot faster than 2G but not as fast as 3G. In planning your wireless application, you can count on 2.5G and/or 3Glite being a reality in most large and mid-sized cities in late 2003. Check the costs carefully because carriers must recover the expense of building the infrastructure, and it won't be cheap. Include in your planning an estimate on data transmitted because the new rates will be based on kilobytes transmitted. Some analysts predict that by 2005 wireless data transmission prices will drop significantly, approaching competitive voice rates. But in the near term, plan for worst-case cost scenarios, which means that even though you can send a lot more data through the air, it will come with a higher price than prior network costs.

Reality Check on Speed Claims

In reviewing the claims of the carriers, throughput speeds of more than 100Kbps are claimed as possible for 3G networks. In reality, the claims about speed usually come from the carriers' marketing departments and represent top-end theoretical speeds that rarely work in practical applications. Why? Because a variety of factors degrade the optimal speed, including the distance between a wireless device and the carrier's wireless base station, any movement of the user such as driving and interference from other transmission sources, limitations on the hardware processing power of the device, and even the strength of the battery!

Actual speeds tend to be around one-half the technology's official rating. The lesson is to test the application in the environment in which it will be used, not just in the lab. On the bright side, even at half the published speeds of 3G we are making a quantum leap forward in wireless data speeds from 14Kbps to more than 100Kbps, making most wireless applications more practical and enjoyable to operate.

Quick Fix: Data Optimization Software

What can you do today to improve the speed of your planned wireless application if you need nationwide coverage? A key problem with moving complex corporate applications to wireless is that these applications were originally designed to run within the enterprise IT system where higher bandwidth and continuous connections prevail. They are agonizingly slow in the wireless world if not on a high-speed wireless network. There is a fix called data optimization software. The price begins around $500,000 and goes up based on the complexity of the applications. By using the optimization software, you can achieve as much as a 25% decrease in end-to-end processing time (what the user experiences). This is like putting a supercharger on your '69 Chevy. It's not cheap but it will run faster. If you have 1,000 users waiting an extra three minutes to process each and every transaction, 365 days a year, the optimization software is a bargain.

Which Protocols Should You Choose Now?

Your choice of a protocol is important at least until the choices narrow. You want to avoid designing a wireless application in a soon-to-be-obsolete or a nonbackward-compatible protocol because it would cost additional time and money to repeat part of the development process. Worse, you could strand several handheld wireless platforms in an obsolete protocol.

The good news is that with continued consolidation and alignment among carriers, there will probably be only two standards competing for the ultimate 3G level. Those two are CDMA2000 (Sprint/Verizon) and WCDMA (almost everyone else including the Europeans and Canadians). Having only two standards in the final 3G playoff round will allow handset makers to manufacture dual-mode devices to accommodate both standards and 90% of the potential 3G markets in the United States. In the longer term, you can relax knowing that every device will eventually be dual mode and will work with every network. By 2005 there may be dual-mode harmonization. For now though, that is definitely not the case.

Between now and nirvana, networks are incrementally upgrading, and you could be caught between technologies. For example, a national company plans to deploy wireless applications to its nationwide dealer network. Looking ahead, it builds the applications based on 3G speeds and the general packet radio service (GPRS) protocol. Once the application is deployed, the company learns that 3G is available only in the major markets and won't be available in the smaller markets (a big deal if you are a Wal-Mart) until 2000-whatever. The smaller markets are still 2G or maybe 2.5G, which runs their application painfully slow. Dealers complain and don't want to use the new system. Lesson: Do your homework on your network geography carefully.

Many consultants will argue that you should develop your wireless strategy to be device agnostic. This means that you would build your processes to work with any device and to accommodate all of the differences among devices and protocols, both now and in the future. To hedge against technological obsolescence amid the rapid proliferation of new devices, this makes great sense, especially in a business-to-consumer (B2C) environment where short messages and simple applications are most common. The middleware software vendors are touting this strategy as a means to combat the wait-and-see attitude that many companies have adopted. I don't completely agree with this approach. Keeping your applications open to morphing to new devices is good planning, but you have to make some device choices in order to deploy a targeted transactional application. The reality is that deploying more complex business applications requires you to choose one or just a few devices in order to fully customize the application to effective use. One-size-fits-all solutions rarely do fit in reality.

What about the Paging Networks?

Using paging networks instead of commercial voice carrier networks to communicate with your wireless users has distinct advantages. First, the paging networks have nearly ubiquitous coverage. Although each paging device works with only one carrier's network, each of the three major carriers in the United States (i.e., Arch Wireless, Metrocall, and Skytel) covers more than 90% of the geographic United States. Even better, most of the paging networks work inside buildings and underground parking structures; there are virtually no holes in the coverage. The paging networks are also very reliable and lower in cost than voice networks. Signals don't degrade through the network chain, and if one transmitter fails, the others operate and the message usually gets through. Combining this reliability with the fact that pagers are always on makes them an almost guaranteed way of getting messages delivered immediately and cheaply (about one-third the cost of telephony networks).

What about Paging Devices?

Pagers have been evolving in features and functions for years. Hardware manufacturers are increasing their screen area, plus adding more processing power and larger keyboards. Pager network providers are upgrading their capabilities for greater throughput and higher speeds. As more devices capable of two-way messaging (like BlackBerry) are introduced, this class of pagers may become even more popular. In the near future, we will see pagers that read faxes, import Web-based information, offer sophisticated time-billing applications, and come with other software tools.

Paging has its downsides, though, including the inability to run popular application software and problems integrating with the enterprise. Some analysts predict the demise of paging networks as consumers come to expect more from wireless computing, video, and the ability to incorporate e-mail with telephony via Smartphone. Personally, I don't buy the all-or-nothing scenario. I believe that in short order the paging hardware and upgraded networks will make compu-paging a powerful and very low-cost alternative for business. Just think of the tier of mobile workers who need a narrow band of information at a very low cost (e.g., pizza delivery drivers, courier services). Cost still matters.

Wireless network technology upgrades are established with anticipated growth continuing every year until 2005. Wireless devices will evolve as well, and not all of them will be backward and forward compatible. Planning the development and deployment of your company's wireless application must include careful consideration of the evolution in network throughput speeds and devices that work with those networks. Why? Because you don't want to frustrate users in smaller markets with a slow and painful wireless experience. You also don't want to strand your investments when the networks are completed.

By now you may be ready to give up and hire a consultant. There is so much to consider and all of it seems to be in flux. But wait, there is more!

WIRELESS DEVICES AND APPLICATIONS

In the near future there will be more mobile devices than personal computers. The landscape in mobile devices is changing quickly, with new and better devices being announced every quarter. There is a lot to know about the devices. This section is designed to help you compare and contrast the various choices presented by your Wireless Action Team, consultants, and vendors.

If a customer-facing wireless deployment is planned, the preferences for devices are largely in the hands of the customer, although you can make recommendations. Make sure you test your application on as many devices as possible to confirm that the anticipated user experience is actually delivered. For employee-facing wireless applications, you have more control over choosing the right device for the business process. Recall that previously, the Wireless Action Team has carefully defined the scope of the application, step by step, whether it's simple e-mail retrieval or more complex order entry. Armed with detailed process information that describes the wireless application, it is then possible to synthesize the user's environment, including the nature and amount of data entry and retrieval. For example, if the wireless application is for security guards who work at night, readability of the screen in the dark is a factor. If it's cold, they often wear gloves, so a tiny touch screen may be difficult to use. Once the wireless application is considered in context with the user, you are ready to evaluate devices.

Consider these 10 factors when selecting the ideal mobile device for the application and the user and their environment:

1. Operating system and processing requirements: Microsoft or Palm
2. Connectivity and coverage

3. Readability (screen)

4. Input mechanism (e.g., keyboard, stylus)

5. Ergo-environmental factors (e.g., case, holster, straps)

6. Processing power, speed, and storage

7. Battery life

8. Startup time

9. Total cost of ownership

10. Technological obsolescence and migration

Better acquainting yourself with these 10 factors will prepare you for the comparative discussions and debates that are inevitable when your Wireless Action Team is reviewing and selecting devices.

Operating System: Microsoft or Palm

Microsoft and Palm are the most common operating system choices for wireless devices. Although there are other operating systems, the most software and integration tools are built around these two brand names. Each has strengths and weaknesses: For graphics-oriented applications, you should choose Microsoft; for forms and text-oriented applications, Palm is very efficient. Palm has been dropping market share for several years, however, as more enterprise-oriented applications become popular. The basic Palm OS cannot run these applications and lacks multitasking, multithreading, graphics, and multimedia capabilities, to name a few. One more major limitation is its lack of connectivity with the LAN standard 802.11 (a and b), which will become a corporate standard very shortly. Palm has responded with acquisitions, new hardware, and a new operating system that makes it more capable of handling enterprise applications. Accordingly, the newer Palm devices should be considered.

Microsoft products have the edge if file processing is needed along with compatibility with the Microsoft desktop. On the downside, the extra capabilities raise the cost of the units and lower their battery life. Finally, the complexity of the wireless application and the need for it to function offline when no corporate system connection is available will influence your choice of operating system. The more you need to keep the device on, the more powerful a processor will be required. The choice of operating system begins with the requirements of the applications but is always linked to the user experience. Choose wisely!

Connectivity and Coverage

Chief Information Officers rank connectivity and coverage issues among their top wireless concerns. They should be yours as well. What if you spend big bucks developing an innovative wireless workforce tool, but workers lose the signal when they are on the outskirts of town or lose the signal entirely when in smaller cities. Network carriers have deployed 3G in large cities and have 2.5G and 3Glite services in middle markets, leaving smaller markets with older service. When my former company was developing our wireless application, the carrier assured us that it had coverage everywhere we required. I asked for a coverage map as proof. They hemmed and hawed and ultimately never furnished it for competitive reasons. Ultimately, the coverage problem limited our potential rollout to only half of our field offices!

Form Factors

Readability of the device screen, the method of inputting data, and the ergonomic issues are together known as form factors.

Readability. Readability is critical to the user's experience. In general, plan on phones having 12 to 20 characters by 3 to 10 lines and PDAs having 30 to 60 characters by 10 to 18 lines. These standards may vary somewhat with different devices between these two ranges. While in the development phase, mock up each screen of your proposed device application, then arrange them in sequence to get a better feel for what the user will see and experience. If the user's work setting is in a warehouse, test the readability in the warehouse. As an alternative, measure the light output in the anticipated work area and then simulate it in a test room using the intended devices. A lighting consultant can help set up the test. Also, remember that a growing portion of the workforce is older, and many of these workers need more light and larger print fonts to read well.

Input Mechanism. The second form factor is the input mechanism on the wireless device. The key question to ask is: How does the user most appropriately enter data and interact with the device? Is the user wearing gloves when data must be entered? Can the application include touch-screen inputs? What about stylus entry? Does the user need a microkey-board (called a Qwerty) for text-intensive data entry? Is a signature impression required?

One way to augment the limitations of phone form factors is to use voice recognition for selected entry. New and powerful software programs are being perfected to make voice recognition a usable tool. Input form factors are the most challenging and are one of the most persistent hindrances to user adoption. Test your choices extensively. Several companies offer services allowing users to speak their requests through a phone then to the Web, where software compresses the voice command and routes it to the company's call center. There the file is decompressed and an operator listens to the request and then executes the instruction or responds. The cost is around one dollar per call. The application for this type of service may be narrow, but it highlights the attractiveness of voice as the primary information-entry mode into a cell phone, which by the way is what the cell phone is best at doing! This solution is especially attractive if access from all types of mobile phones or land-based phones is a requirement.

Ergonomic and Environmental Factors. The last of the three form factor issues involves how the wireless device will be held, carried, holstered, slung, or stored, combined with the environmental elements such as exposure to dust, rain, snow, heat, or chemicals. How rugged does the device have to be? In all cases, plan for the worst. Are there any other special factors to consider? A handheld used on an oil platform must be quite different from one used in a shoe store. If your company has especially harsh environments, do some more homework. Ask the manufacturers for the names of companies using their devices in similar environments and check out the references. Ask them for specifics about their devices. When all the data are in and a few devices are selected, give them to a few users to test and see how they hold up. Test and test again.

Processing Power, Speed, and Storage

Processing power, speed, and storage (onboard memory capability) are the three key performance factors. Device processing power varies widely, and a little more processing power can go a long way. To benchmark, use as a rule of thumb that the processing power in the best PDA is only about one-tenth as powerful as a current desktop computer. PDA screens are also about one-tenth the size of a desktop monitor. This is counterintuitive for the PC generation because for years we have enjoyed more power (faster processing), more memory, more graphics, more everything. Pocket PCs are better, but still less than the desktop, at least for near term. The challenge comes when attempting to run familiar enterprise applications that rely on powerful PCs, servers, and wired networks, with smaller, less powerful handhelds.

Device performance is rapidly improving and evolving. The old standby ARM processor is eclipsed by the XScale processor from Intel, and that processor will eventually be eclipsed too. Storage capacity is moving up quickly. Whatever the current technology, the wireless computing application must be planned carefully and must be lean in order to squeeze every advantage from the processor and storage parameters. Do some homework and find out when selected manufacturers are planning to upgrade their processors. If the planned wireless application runs primarily on the device (thick client), ask your IT staff to calculate the impact of the new processor on the user experience. It may be worth the wait.

Battery Life

Additional capabilities such as color screens, Internet browsers, and video all increase a device's power consumption. Nothing saps wireless user enthusiasm like running out of power between charges. Enhancements in battery technology, such as lithium-ion batteries, have taken laptop power life to a level only imagined in the late 1980s, but the technology has yet to be successfully extended to the smallest of devices—phones and super PDAs. Rayovac announced a breakthrough in rechargeable batteries with a new nickel metal hydride (NiMH) battery that lasts four times longer than current lithium-ion batteries, charges faster, and takes up to 1,000 charges. They promise the battery will be commercially available in 2004.

We will need it and more because the current battery technology is unlikely to keep pace with the power demands of the next generation of feature-rich and power-hungry wireless devices.

Power will be a major limiting factor on the practical capabilities and functionalities of future devices. Ask the device manufacturers for power utilization and battery life performance metrics. Use these to compare and contrast device choices. For service and field personnel, have a team member ride with them for a day to measure how much charge time may be possible. Factor the charge time projections into your battery life calculations and include the cost of batteries, rechargers, and disposal into your cost calculations.

Startup Time

How fast the handheld boots up and connects to the Internet or network is a productivity and user satisfaction issue. Faster is better. Many of the newer units are always-connected devices, meaning they can receive e-mail or alerts almost all of the time. Not withstanding the claim, take the device through the paces of powering up, loading, and then using the planned application. Use a stopwatch to measure the startup times and compare the selected devices. Add startup time to end-to-end process measurements where appropriate.

Total Cost of Device Ownership

The total cost of device ownership (TCO) is important in an employee wireless deployment. It comprises the following elements:

- Initial acquisition price of the device, accessories, and software amortized over a 12- to 24-month period, depending on the environment
- Licensing costs (if any) per device
- Cost of charging equipment and batteries
- Cost of carrying cases and harnesses
- Incidence of failures and the cost to fix and redeploy
- Spare devices required based on anticipated failure rates

○ Spare chargers, cases, and batteries based on estimated loss and damage

The total cost of ownership (TCO) elements are often unique to an application, and your team can probably come up with a few more cost elements to put an accurate number on the annual cost of owning and operating a wireless device. The differences between devices may be larger than you would imagine. As an example, experts estimate the first year total cost of ownership for a laptop is about $9,600, a wireless-enabled PDA is between $3,000 and $4,000, and a BlackBerry is about $900. Choosing one does not necessarily eliminate the other two, but where possible, it allows the deployment of a lower-cost or more efficient tool. Once you've estimated the TCO per device, multiply that figure by your workforce to get the comprehensive number. Use this estimate to compare device choices and negotiate with suppliers.

Technological Obsolescence

The technological lifespan of mobile devices is about half that of a PC or about 12 to 24 months, depending on when the purchase is made. Similar to PCs, when a purchase is made immediately after a new processor development or other significant enhancement, the technological life is greater. I learned one aspect of this lesson from early experience. We had placed an order for 500 Palm VIIs and were excited about our planned wireless rollout. I woke up one morning to news that Palm was launching the Palm VII X. The X meant more memory—four times as much memory. Our application was significantly enhanced by the extra memory. With little time to spare, I cancelled the order and placed a new order for the X version. Why hadn't the corporate retailer notified me that the new version was coming in just a few weeks from my order date? Let the buyer beware! Direct your Wireless Action Team to do some homework and plot the device manufacturers' product release plans. In doing so, you will avoid being stranded with a brand-new but inferior model.

Benchmarking and ranking your prefinal selections against these 10 factors will help you decide which devices work best with your wireless application. Making the final selection may involve a series of tradeoffs:

cost versus features, expandability versus size and weight, custom software that is exactly what is required versus off-the-shelf software that covers most of what you need.

Also keep in mind that new categories of wireless computing devices continue to evolve. Smartphones and wireless Web tablets are two promising categories. Both have made the transition from cool toy to serious business tool, and they should be considered as viable options for the enterprise or customers.

Smart Phones

Like Batman with his utility belt, today's totally connected worker could have a cell phone, pager, and PDA with a modem to access the Internet. Multiple devices are inconvenient and take up space, not to mention the hassle of keeping all of the batteries charged. The alternative is a single device with all of the features we need and want—the smart phone. Smart phones may become the best alternative for the wireless professional workforce and customers. This category of devices has the potential to rise quickly in popularity, especially as users grow accustomed to the convenience and functionality of having one device to meet all of their needs. Research firm IDC estimated that the market for smart phones will grow from 12.9 million units shipped in 2001 to more than 60 million in 2004.

A workable smart phone must first be a really good phone, then second a pretty good PDA. Palm-based PDAs led the smart phone pack initially but were quickly followed by several integrated devices using windows-based offerings and most important, the Microsoft Stinger Smartphone. All of them synchronize to Microsoft's Outlook. Running on the newer version of CE, the Stinger also allows users to view e-mail attachments, such as Word documents and Excel spreadsheets. Of course, there are concerns like shorter battery life and the learning curve for managing the phone while operating the PDA, but new technology in batteries and the advent of clip-on microphones and even clip-on visual displays make smart phones a smart choice.

Wireless Tablet PCs

Usability is the main barrier to extending more complex applications to smaller and smaller devices. Just because you may be able to use an Excel spreadsheet on your wireless phone doesn't mean you would actually enjoy using it, much less find it practical. The solution may be a product halfway between handhelds and laptops—the wireless Web tablet or tablet PC. This family of products provides a larger screen than even the largest handsets but is smaller and lighter than the most compact laptop. This class of devices is especially interesting because wireless business users often need to review more data than a few small screens, as well as complete service order forms or view technical documents.

How information is entered is a key developing point in tablet PCs. One camp believes in completely eliminating the keyboard and substituting a stylus (like a pen with no ink). This approach is more natural (and is being called natural computing), but it is not without problems, such as imperfect handwriting recognition software. The other approach is to compress and slightly reorganize the standard keyboard to fit a smaller device. Although still evolving, this group of products is promising as a compromise solution between small PDAs and laptops. Tablet PCs with wireless capabilities may offer a good solution to more complex mobile computing challenges.

SECURITY CONCERNS IN MOBILE DEPLOYMENTS

The rapid development and widespread use of the Internet, wireless devices, and WLANs in our personal and business lives has far outpaced our application of security technologies. Security is often the last item considered in a wireless deployment, even though a variety of off-the-shelf security applications are available. Salespeople are lined up in the lobby to make presentations on special firewalls, intrusion detection systems, wireless security, and vulnerability studies. When discussing security, there is a "sky is falling" hype that can invite overspending. Like everything else in wireless, it requires taking a realistic approach.

IT departments are not enthusiastic about wireless workers for a lot of good reasons. A remote wireless workforce makes it difficult to maintain clear control over access to systems. Having hundreds or thousands of remote wireless workers with access to the mother ship is like locking the front door to the house but leaving hundreds of windows open. There is no question that wireless worker devices are the weakest security link in the overall network security.

A wireless security consulting team working for Crowe Chizek, a public accounting and consulting firm, conducted a three-day survey of WLAN security in Atlanta in summer 2002. The investigating team drove around Atlanta probing wireless local area networks that leaked out to the public street. Without actually penetrating each WLAN, the security team identified network access points when the unsuspecting network provided an unprotected handshake to the security firm's wireless laptop, logging more than 1,000 networks without adequate security protocols in just three days. The security team detected a lack of even default encryption protection in two-thirds of the systems, an indicator that no additional protection existed.

Even with default encryption protections, no network is ever completely secure, at least to someone who wants the information badly enough; however, with some simple and somewhat inexpensive steps, you can make it very expensive to breach your network. There are hardware and software firewalls available for every size business. Each claims that it is the best at identifying worms, viruses, Trojan horses, and other nasty beasts, so where should you begin with security?

Security for wireless should be based on specific business objectives filtered through straightforward risk management thinking. Technology is *not* the place to begin your thinking. The recommended steps for building your wireless security program are as follows:

1. Use the process maps developed in the beginning of your wireless planning to identify potential security compromise points.

2. What if a device is stolen—is the device password protected? Are the network logins stored on the device? Do you have a device management system that monitors active and authorized devices?

3. What protections are built in at each layer of access—the WLAN? the corporate firewall? the application server? databases?

4. Build your policy statements around the business rules and operational protocols that make sense in addressing the security compromise points.

5. Assign personnel to have ownership of the business rules, operational protocols, and technology tools. Write the procedures and standards to blend all of the security program elements.

6. Explore technology solutions that economize or optimize the application of security controls in harmony with your business rules and operational protocols. (Don't spend more than the value of the information!)

Security is critical during transmission, at the remote device, and at the wireless gateway or WLAN entry point. Encryption is the first level of security and is applied at all three of these points. Depending on the importance of the transmission, you can use the out-of-the-box encryption that comes with each application or you can buy more sophisticated software. Authentication on the device can be accomplished via password with time-out features requiring reentry after being idle for so many minutes. Most PDAs and some mobile phones have this capability built into them. Finally, you will probably have to purchase some software to guard your wireless gateway or WLAN point, the entry point to your business system, and then coordinate that software with your enterprise systems and the wireless transmission and device security. Once again, the more you spend, the more you get (usually), but the most important question is—what do you really need?

Hacking individual Web transactions via wireless is far less attractive to hackers than breaking into a server to get thousands of credit card numbers. Rarely will anybody steal information that they can get legally, but, industrial espionage is not out of the question. Notes from potential merger meetings, new product development notes, and customer-pricing profiles are all valuable to competitors. Thus competitors or others intercepting e-mail is a potential problem and can be addressed by creating more secure tunnels for the e-mail streams to travel. Unauthorized access to company databases is another potential security liability, especially when passwords to these databases are stored on the wireless device.

One solution is building protocols that require users to enter the password each time they use the device. Prompt cancellation of passwords for ex-employees or for lost or stolen mobile devices should also be a priority. Security firm Pointsec carried out a survey of London cabs and found that in just six months around 2,900 laptops, 1,300 PDAs, and more than 62,000 mobile phones were left in London cabs. So much for airtight device security![1]

Another alternative to beefing up wireless security is to set up a virtual private network (VPN). A VPN extends your local wireless network across the Internet through a protective tunnel from laptop to server. VPNs are not cheap and currently don't work with PDAs and phones.

To determine the level of security your applications require, ask yourself what the financial impact would be if a particular element were compromised or a database was erased. Spend accordingly to protect the most important information and systems.

WIRELESS OPERATIONS SUPPORT

Once you have invested hard work and expense in launching your wireless application, it is important to support the wireless users. When the wireless program is deployed, you'll need to publish a user guide, FAQs, and information on who to call to get support. Your help desk or customer service group will have to be trained on the new product and have a clear understanding of the role played by the wireless application and the wireless device.

A key support issue will be measuring and managing degradations or outages in the network. One approach is to create in-house processes that test the log-on times and run times for your wireless applications on a systematic basis. Your wireless carrier may also offer some reporting capabilities, but beware of the positive spin they will give to their numbers. Another option is to hire a service company to monitor and test your network and applications. This service costs a little more, but the reports are comprehensive and objective.

The most advanced service companies will mirror actual end-user experiences by setting up testing protocols with all of the wireless devices that can access your system. The devices, which are controlled by the monitoring company's software, continually attempt to run your applications and send and receive data. The third-party service company produces a report on end-user performance as well as immediate alerts for downtime or other problems. Although these reports are point-in-time oriented, they may be your best method for mapping trends.

If your IT group and the various third-party providers can agree on a set of metrics for performance, you would do well to tie the performance metrics to the original purchase price (partial refunds for failure to meet promised performance) and, more important, to the ongoing monthly fees for the network and to software maintenance fees. Linking performance to payments can be described in a document called a Service Level Agreement (SLA). Although the major carriers claim to offer SLAs, when you look at the fine print, many lack real guarantees. Because of fluctuating bandwidth availability, it is very difficult for carriers to make unconditional guarantees. What they will guarantee is the speed between points-of-presence or key nodes on their network. They will not guarantee an end-to-end speed, which would more likely reflect a user's real experience. Negotiate aggressively for rate reductions in cases of substandard performance, reduced uptime, or coverage expansion milestone failures.

REFLECTING ON THE CHAPTER

As you can now appreciate, extending your company's enterprisewide system to wireless is a huge undertaking. Each of the four layers must be addressed for a fully integrated and synchronized application. You will need your in-house IT teams to lead the technical evaluations of software, networks, and devices. In turn, they will need support from executives to champion the effort through the inevitable setbacks and surprises. A timely, on-budget wireless deployment is achievable. The next chapter summarizes the lessons learned by companies who blazed the trail in enterprise wireless. Learn from the best!

ENDNOTE

1. "Companies are afraid of losing mobile devices for security risks," *GSMBOX.com* (2002), online at *http://uk.gsmbox.com/news/mobile_news*.

8

Top Ten Wireless Lessons Learned

This chapter has some good advice. Some of the advice originates from my own experiences, but more important, most of it comes from conversations or readings taken from the First Practice companies' wireless champions. Of all the lessons that can be taken from the First Practice 50, these are the ones that really count. The top ten wireless lessons learned are as follows:

1. Target a business goal that can be achieved with wireless and yields a favorable ROI.
2. Keep it simple and start small.
3. Use a maximum ROI of 12 months. Invest incrementally and prove up the ROI as you progress.
4. Commit to build inhouse wireless expertise.
5. Anticipate rapid evolutions in wireless technology planning and budgeting.
6. Address the enterprise architecture challenge.
7. Get serious about wireless security.
8. Assemble a Wireless Action Team.
9. Include the user from the beginning of the project.
10. Start your wireless initiative today.

LESSON 1: Target a business goal that can be achieved with wireless and yields a favorable ROI.

Technology is a means to an end—more productive and competitive business operations. Accordingly, there is no need to deploy wireless unless a specific business process is improved and it complies with one of my favorite laws of technology economics: Technology must add more value than it costs to acquire, deploy, and operate.

This seems intuitive, but in the past technology was deployed in many companies (not all) simply because it could incrementally improve a process or improve the timeliness of information delivery by some degree, not because in doing so the cost of operations would decrease. Don't be afraid to say "not now" to wireless when the deployment doesn't clearly meet this standard.

Users—as well as competitors—are an excellent source for ideas about process improvements using wireless or other technology.

LESSON 2: Keep it simple and start small.

This lesson is contributed by Will Taylor, Senior Engineer at Averitt Express. Keeping it simple means minimizing the number of screens and graphics. Make do with as little information on the handheld unit as possible while placing other information on a barcode or other automated entry mechanism. Stick with simple text if possible. Minimize the number of menus and their contents. Keep the number of steps in any one process to two or three. Only provide information that is directly relevant—the most important information. Keep the information targeted, concise, and easy to access.

This lesson was also endorsed by Avera McKennan, the home health care company. Its health care technicians do not have to enter a lot of information about their patients into their handhelds via keyboard entry. Instead, they simply scroll through menu choices and highlight their options. The software application compiles the choices into a patient history that is easy to view. Make it easy to use and it will be used. Keep it simple.

LESSON 3: Use a maximum ROI of 12 months. Invest incrementally and prove up the ROI as you progress.

Tim Van de Mewre of Associated Food Stores told his wireless vendors upfront that he needed a guaranteed ROI in 9 months. Tim believes that the ROI must prove up in less than 12 months. He commented, "You don't want to depreciate technology investments out over the standard 3 to 5 years; that cannot be the rule of thumb with technology because there's going to be something bigger and better and you don't want to be tied down to your technology investments as they become obsolete."

As the wireless application moves toward accessing corporate data and working in more complex processes, the costs and time to develop and implement the application increase significantly. At this point, more costly decisions are made with regard to architecture and middleware. Future scalability is weighed against cost and time to market. Take the extra time to work through the ROI analysis in as great a detail as possible. Divide the investment into functional phases, if possible, with ROI calculated to a maximum of 12 months, as well as verification of ROI at the completion of each phase. Begin with the smallest possible phase that makes sense, confirming usability, performance, and ROI before beginning the next phase.

Lastly, recall the glory (or gory) days of the Internet boom. The rush-to-build bug had infected corporate America, triggering a software feeding frenzy. Companies rushed to deploy Web commerce by buying highly touted software at a premium price. In the rush, companies substituted their usual careful analysis with a risk-based approach. This resulted in buying so-called best-of-breed applications, usually the most expensive, and then cobbling them together. Although the recent lesson is still relatively fresh, some companies may repeat the mistake by not taking their time with wireless deployments, learning the space, testing their own skills, comparing software products, and carefully negotiating agreements. Overpaying or overbuying can kill your ROI, so plan carefully and buy right.

LESSON 4: Commit to build inhouse wireless expertise.

Wireless should become an integral part of your IT infrastructure—forever. Part of committing to wireless for the long haul is building in-house wireless knowledge and expertise. To build in-house capabilities, your company's IT group should create as much code and integration components as possible, given the size and operating budget of the department. In-house IT ownership of the code-level details of wireless allows the IT team to be more informed and candid about the components which need to be acquired from software vendors or coached by a consultant. Committing to wireless for the long haul and to the in-house talent to support it will in the long term, after all the inevitable changes and evolutions, keep you in control and save money.

LESSON 5: Anticipate rapid evolutions in wireless technology planning and budgeting.

If there is one certainty, it is the certainty of change, especially with wireless. Rapid changes in technology infrastructure make most executives uncomfortable. No company wants to be stuck holding the obsolete technology bag. Wireless, like a lot of other technologies, is not something you buy one year and then you are finished. There will be multiple waves of investments supporting iterations of wireless deployments and upgrades. It goes on, and on, and on—hopefully with a favorable ROI. Get comfortable with it because the more you know the more you can plan. With a little planning you won't be surprised when user preferences in hardware change, more powerful hardware platforms emerge, and bandwidth and data transmission speeds increase. You will have incorporated them into your planning and budgeting processes. Plan and manage technological obsolescence issues or they will manage you.

LESSON 6: Address the enterprise architecture challenge.

It is good to begin with wireless applications like e-mail, calendars, contact lists, and phone directories, but when you are ready to move to more complex applications requiring interaction with corporate databases and systems, beware of deploying the application without addressing the changes to the enterprisewide architecture that will allow scalability and flexibility in the future. This doesn't mean that a one-off application is necessarily bad because often you will need a successful application to prove up continued and larger-scale investments. The lesson learned is that once you are convinced wireless has a place in your company, you should plan the future accordingly. The details behind this lesson are technical in some ways but mostly common sense. If you were building a house, small and conservative, but knew you were going to add rooms some time in the future, you would plan accordingly by extending the foundation, making sure there was connectivity to plumbing and electrical, and more important you would orient the design of the house in considering the future add-ons. Approaching the enterprisewide architecture for wireless is similar. To do it right you need to plan ahead and make the changes in system architecture, primarily middleware, to accommodate the current and future iterations of wireless applications accessing multiple databases and applications throughout the enterprise. When you are certain that wireless has a future in your company, face the enterprise architecture challenge.

LESSON 7: Get serious about wireless security.

As the number of wireless devices and WLANs grows exponentially, the number of unauthorized pathways of entry into the corporate enterprise grows as well. Basically, wireless devices and WLANs are inherently insecure. To avoid nasty surprises and financial losses, get serious about security. Encourage your IT team to list security as a top-of-mind issue for every project. Insist on and fund security training to build security competency. Construct layers of security to protect your network and your data, including authentication, encryption, software and hardware firewalls, aggressive password procedures, and employee access management programs. Test your security by assigning special in-house teams to attack your security and reward those who discover weaknesses. Get free software programs like NetStumbler that allow you to test your own WLAN security. An ounce of prevention is worth a megabyte of secured data.

LESSON 8: Assemble a Wireless Action Team.

Forming an in-house cross-functional team to analyze and plan your wireless strategy and implementation is critical. Whether the resources are dedicated exclusively to wireless depends on your scale and the scope of the project; but regardless, you need a team. Winn Stephenson, senior vice president of IT for FedEx, said it clearly, "You need an in-house team to sort through all the variables."[1] This is true for small and large companies. For customer-facing applications, the customer must be on the team from the beginning. Assembling the wireless action team and accommodating customers and nontechnical members is time-consuming, but it is an essential step to ensure against creating an application that does not consider all of the variables from billing, inventory, logistics, and legal, to security, marketing, human resources, and so on.

LESSON 9: Include the user from the beginning of the project.

The best source of information for improving or correcting a piece of technology are the users—either customers or employees. Including them in the process is a beginning point, but you should go much further. You have to get really close. Establish a collaborative environment where critiquing the new process or technology or feature is not just allowed, it is rewarded. Charlie Feld, former CIO of Frito-Lay, described its deployment of new handheld computers to the small handpicked pilot group.[2] Feld showered the pilot group with attention and encouragement while effectively soliciting their feedback and addressing concerns immediately.

When piloting a wireless application with a customer group, I provided free breakfast and feedback sessions every week and visited customers at their worksites, asking questions and soliciting their impressions of how they would change the application for the better, nodding yes and writing down every word. Staying close to the user does take time and it does have a cost, but it protects you from failure far better than any consultant or software application.

LESSON 10: Start your wireless initiative today.

Having a wait-and-see approach to new technology is not altogether a bad strategy. I mentioned earlier one of my former companies' "second mouse gets the cheese" approach, where other companies act as the guinea pigs for testing new software or act first in investing where customer demand was unproven; however, somewhere after the other companies have done some pioneering but before all of your competitors have wireless, you need to get started with wireless.

Why? Because the learning curve in developing an organizational competency and savvy with wireless does not happen overnight. Getting started today will begin the process of building knowledge equity and a vision for what's possible in your IT, operations, and sales teams. The best way to avoid overspending in any area is to not be in a hurry and to have inhouse expertise.

When Fidelity Chief Wireless Officer Joseph Ferra would attend management meetings and discuss his vision for wireless, the other executives would ask if Fidelity should wait for a faster network or cheaper devices.

Ferra always answered "no." He believed that starting with a simple application today, piloting carefully to prove up demand, and then investing incrementally was the right approach. Today Fidelity is at the top of the list of investment companies offering comprehensive wireless services to their clients. This lesson sends the message that you should get started today—keep it simple and even small—but get started. Phillip Redman, a research director at Gartner, declared, "CIOs need to start planning for wireless. Those who wait will be left behind."[3]

REFLECTING ON THE CHAPTER

The top ten lessons are deceptively straightforward but tougher to follow strictly during a complex project. You and your team bring a wealth of lessons learned to the wireless development challenge, mainly your common sense. Good luck on your wireless project!

ENDNOTES

1. "Survival Tips from the Pioneers," *CIO* (March 15, 2001). Access at *www.CIO.com*.
2. *Ibid.*
3. Tom Field, "Five Uneasy Pieces: Beneath the Hype of Hot Technologies," *CIO* (November 2000).

9

Wireless Forces of Change

At this point in the book, you have read many examples of companies using wireless applications in some manner. Still skeptical? If you are, that is somewhat understandable because there have been some significant barriers to the mainstream business use of wireless, such as an uncertain ROI, questionable network speeds, and restrictive device usability. One of the most important messages of this book is that those barriers are fading, and fading fast. Three powerful forces of change or *tipping factors* are accelerating and driving the practical deployment of wireless in business. They are:

1. *The need to keep the ever-growing mobile workforce productive.* There are more mobile workers than ever, and their numbers are rapidly increasing. Wireless computing extends the functionality of the enterprise system to the mobile worker, transforming him or her into the new electronic worker, allowing timely and efficient process improvements never before attainable. Companies see wireless as a way to cut costs and improve productivity, and there are dozens of examples to help construct realistic ROIs.

2. *Huge improvements in network performance and device functionality together with a critical mass of deployed WLANs will make wireless easier to use than ever before.* Networks are finally leaping from slow to speedy, and devices are improving their form factors and processing power significantly while adding WLAN connectivity to connect with millions of deployed WLAN nodes. We are rapidly moving from toys to tools.

3. *Always-on connections for voice and data are steadily moving from wants to must-haves.* Customers have experienced the convenience and

control of the Internet and wireless is the logical extension to access the Internet, anytime and anywhere. Our society as a whole is rapidly adapting to and expecting an always-connected environment—the wireless Internet and the real-time information it provides.

This book is about facts and examples instead of conjecture and hype, so let's examine some data that support these assertions, beginning with the trend toward connecting the mobile worker to the enterprise.

COMPANIES WANT THEIR GROWING MOBILE WORKFORCE CONNECTED AND PRODUCTIVE

At the center of the rapid growth in wireless usage is the need for executives, salespeople, and other mobile workers to maintain a real-time linkage to the hyper-paced business world. Eight out of 10 companies considering a wireless deployment would deploy to their own employees first, according to a 2001 survey of 150 companies by IDC. Workers come first for two reasons. First, companies don't want to test out a new technology on their customers and, more important, the mobile workforce is the fastest-growing segment of the workforce, increasing by 9% annually to reach 55.4 million by 2004, up from 39.2 million in 2000.

Most businesses believe that deploying wireless to the general workforce will boost productivity and cut operating costs. As early as 2001, the Economist Intelligence Unit surveyed 172 business executives from the United States and Europe about their thoughts and future plans for wireless.[1] More than 60% believe that increasing the efficiency of business processes is a highly important factor/driver in their wireless investment plans. The study further indicates that over a three-year period, executives expect that wireless could deliver productivity improvements and/or cost savings ranging between 12% and 16%. Most said they believe the improvements would come in steps, and to achieve the savings their companies must develop a deeper understanding of how to use mobile computing to their advantage. For most businesses, improving worker productivity by even half of what these executives estimate would be significant.

There is no shortage of research supporting the direction companies are taking to capture the productivity potential of wireless. A mid-2001

Forrester Group study of the Fortune 2,500 (US) revealed that 40% have equipped or were in the process of equipping their workforces with some form of wireless tools, and another 30% were "considering" rolling out wireless to their workforces.[2] Overall, the survey results yield a solid 70% participation rate, a figure that competitive companies can't afford to overlook. As a conservative qualifier, we don't know the degree to which wireless has penetrated the workforces of these companies. It may be that executives, field salespeople, or some other isolated group have wireless capabilities or simply that the proof-of-concept process is underway. Regardless, the direction is clear.

Most companies will start small with wireless, testing with key groups, learning, testing again, and then rolling out to larger groups of workers. When the early wireless pioneer Frito Lay committed to rolling out a wireless order management tool to its 10,000 field reps, it handpicked a group of 50 reps to work closely with it to perfect the application. Once all of the bugs were worked out and the technology and processes were proven, the rollout went smoothly. Over time the employee culture will become even more acclimated to the technology. You will probably notice that almost 70% of the 50 First Practice company examples in this book deployed first to workers in hopes of increasing productivity and streamlining business processes.

Equitable Life Insurance purchased 3,650 Compaq Ipaq pocket PCs for its salesforce to be able to access the corporate network while on the road. Equitable salespeople can now access broker records, activity reports, and transaction histories, and synchronize their appointment calendars, contacts, and to-do lists with their corporate network. Thousands of previously disconnected field agents are now connected in real time—all the time. The technology also provides a means to better measure and manage the performance of Equitable's salesforce. Will others in the insurance industry follow? You can bet on it.

It's more than just a productivity improvement; real-time connected employees have an edge with customers. A real estate agent in a hot market would benefit from knowing as soon as possible when a new property is listed. Those who have the information first often score the sale. It won't happen overnight, but companies are clearly moving forward in the deployment of wireless to the workforce, and it would be a mistake to underestimate the competitive drivers of wireless. Wireless computing is this decade's weapon of choice for employee productivity and competitive advantage.

HUGE IMPROVEMENTS IN NETWORK AND DEVICE PERFORMANCE WILL MAKE WIRELESS EASIER TO USE

One reason companies have hesitated in deploying wireless to workers on a broader scale is the frustration factor. If you want to keep your workers happy, you don't force them to use new tools that even when mastered, take longer to do the job! Wireless may (and computing in general certainly does) have that poor reputation. To date, two technological constraints are the culprits behind the frustration factor: slow network speeds and limited functionality of handheld devices. These limitations and the accompanying uncertainties are on the verge of being removed from the adoption equation. Networks are leaping from slow to speedy as the next generation of data infrastructure comes online, and devices are rapidly evolving, and improving their form factors and processing power significantly. Achieving this next level of functionality and performance with networks and devices will be another "tipping factor" toward broad wireless utilization by business.

WIRELESS NETWORKS GET A SUPERCHARGER

Network bandwidth and speed affect the amount of data that can be transmitted wirelessly. Slower speeds mean larger packets of data for image files or spreadsheets must be broken into parcels and moved one at a time over the network, then reassembled to form the image or file on the mobile device. This takes valuable time and frustrates users for whom the wait can be agonizing. In the past, we haven't had the wireless speeds necessary to operate on the Web without frustrating the user. Add to that the time necessary to logon, and the wireless Internet session is just too inefficient for more than simple applications. Whether customers or employees are the users of the wireless application, the elapsed time to use it must be at least the same as or faster than the nonwireless process.

In 2003 we should see the completion of key deployments of new telecommunications and data infrastructure that will dramatically increase data transmission throughput (speeds). When it used to take 10 minutes to download a data file to a PDA, it will now take 10 seconds. Even more important is the promise that the new network protocols will allow an always-on functionality for data—no delays in connecting to the network.

That means quicker access and response times. For now, the most important point is that the networks will be supercharged, and virtually all wireless applications will run faster. Users will begin to smile, or at least complain less!

At the same time the data and telephony networks achieve high-speed capability, WLANs will be everywhere imaginable. Why? The lower cost and improved performance of WLAN equipment will pave the way for tens of millions of new WLAN nodes (lily pads) to be deployed. Small businesses and retail establishments will lead the way, with WLANs becoming the standard for network deployment. PDAs, laptops, and even phones will be configured to connect to any WLAN and establish a high-speed link. Gartner reported that global WLAN shipments were 15.5 million units in 2002, and in 2003 it is forecast to be 26.5 million units. By 2004, 31 percent of all mobile PCs are expected to have integrated WLAN features, up from 10 percent in 2002.[3] The impact of WLANs may be as significant as the PC itself.

MORE MOBILE DEVICES AND BETTER APPLICATION SOFTWARE

Another barrier to widespread business adoption so far has been form factor and usability of wireless devices. Phones are really good at making phone calls but pretty painful to use as Web browsers. Except for very basic selections and interactions, conducting wireless computing over a cell phone is relatively difficult. The screen is too small and the telephone keypad is difficult to use for alpha character entry. The exception has been where narrow software applications were written for a specific need that had lower demands for large amounts of data entry. At Carolina Floors, a North Carolina floor contractor, job foremen enter the payroll hours for their construction crews on their Nextel phones with an hours worked application, and then forward the information to the head office for processing. Instead of calling in each worker's time each week, and having it entered into the accounting system where payroll was calculated, the new process allows the job foremen to enter the hours, then forward them electronically to their office, where it is uploaded into the accounting system. The new process has completely eliminated the reprinting of erroneously calculated payroll checks as well as reduced the cycle time of payroll processing from five hours to one hour. The application also allows CEO Jeff

Rogers to have more visibility to labor distribution data and better track job progress. This simple application was Carolina Floors' first attempt at using wireless to improve operations. CEO Jeff Rogers described installing a wireless application as "the best money I've ever spent."[4] Phones work adequately for simple applications, but what about more complex applications using PDAs?

A huge leap in usability was made when PDAs went wireless. More PDAs than ever have been enabled for wireless access to the Web, screen sizes have been doubled, and options for input have increased. Combining the PDA capabilities with targeted software applications is working. For example, ServiceMaster utilizes wireless-enabled PDAs at 20 Greyhound locations to record the results of inspections on the cleanliness of Greyhound buses. The company dropped its paper-based system and issued Palm Pilots to its 50 inspectors. The inspectors now report daily instead of weekly. Inspectors enter data into their Palms and then upload to the ServiceMaster server, which compiles the data and produces reports for Greyhound. Greyhound can correct problems faster by getting the information quicker.

A new batch of business-oriented wireless PDAs have been introduced, with more coming. New off-the-shelf devices by Symbol, NEC, and others target particular industry niches, and combining them with the appropriate software can provide a highly usable tool. Office Depot, a leading big-box retailer of office supplies, deployed Symbol wireless handheld units to track its delivery fleet and to capture customer signatures on delivery of office supplies. With thousands of trucks delivering to tens of thousands of customers, Office Depot needed a way to speed the billing process. Unable to bill customers until the delivery ticket was signed, the driver returned the ticket to the local store, and then the shipment was billed via key entry into a billing system to invoice the customer. With the wireless devices, the signature is captured in electronic form and routed in real time to the Office Depot system, which immediately generates the invoice.

New devices are continuously introduced with an impressive number of cumulative improvements. New and improved wireless phones with larger screens, scroll keys, and more powerful processors; new and improved PDA smart phone combination devices; new and more powerful wireless-capable PDAs; and new tablet and clipboard PCs are starting to catch on.

Regardless of how many devices actually live up to all of vendors' claims, it is undeniable that the devices are getting better and the number of choices is increasing. The bottom line is that devices will get good enough and people will adapt enough to incorporate them into mainstream use in their personal and business lives. That's not all, though; for those of us a little past the millennial generation, we can look forward to better handwriting recognition software breaking through to usability as well as very accurate speech recognition software hitting the market in 2003. These will further reduce the form factor barriers to wireless and broaden the usability to a larger audience and more industries.

WIRELESS USERS FOR BOTH VOICE AND DATA ARE GROWING

Last but not least, wireless usage for both voice and data will move from novelty to norm, growing at an amazing pace as society broadly acclimates and adopts the wireless Internet. Examining the growth in wireless user numbers, it probably would not surprise you that the fastest-growing access to the Internet is not through the PC, but through wireless devices. In 2000, there were fewer than 2 million users of the wireless Web in the United States and only 10 million in Europe (as per a study by the Economist Intelligence Unit).[5] In 2002 the number was estimated at 7 million in the United States and by 2005 the number of U.S. wireless Web users is estimated to reach more than 100 million, according to three research organizations, Ovum, Gartner, and Forrester. Even if scaled down, the growth curve is significant.

The aggressive predictions are based on several assumptions that include (1) the continued offering and upgrading of wireless Internet access as the upgraded telecommunications networks offer 2.5G and 3G in more cities; (2) the anticipated appeal of commerce by phone; (3) and the future introduction of streaming audio and video content in 2003. Most of these studies also cite improvements in software and video as well as lower costs in devices and networks.

The numbers predicted point to a 7- to 10-fold increase in the United States. Unfortunately, many Internet or telephony-related studies tend to "hockey stick" the trends and curves up higher and faster than what actually happens. Although it is easy to criticize the research organizations for the variations, the forecasts are valuable in helping to validate the anticipated surge in wireless Internet users, even if the forecasts are widely off. Filtering the data with a conservative interpretation of the forecasts, it is still certain that wireless users overall are predicted to grow substantially, and along with that will be their use of the Internet.

You may react with the comment, "That's mostly consumers," and you may be right to a degree. It is very important to compartmentalize consumer wireless trends, applications, and predictions from the business use of wireless; however, it is also important to understand and briefly discuss the consumer side of wireless in order to understand the related business-to-business (B2B) issues because wireless is a technology led by perceived consumer demand.

Despite the lingering limitations, more consumer-based services are in the works that will incorporate the use of wireless PDAs for transactions and information retrieval. Coming soon are services that support using a PDA as a passport to an ATM, gas pump, or for other point-of-sale transaction. Other consumer-type services such as direction finders, restaurant guides, travel updates, sales finders, and price comparators are being offered and may take hold. As on the business side, consumer wireless is nearing an important crossroads where the upgrading of the telecommunications infrastructure, together with new, more powerful devices and some new content, will begin to meet the high expectations of consumers. Although this book is focused primarily on B2B uses of wireless, many of the ideas may apply if you desire to build a B2C model. It is also important to note that many of the new services and applications that are used in B2B were developed with the hope of being part of the killer app catalyst for a breakthrough in the consumer markets. The uncertainty and confusion in consumer wireless may tempt you to discount the value of wireless computing for business. Don't do it!

SOCIALIZATION OF WIRELESS

There is an intriguing aspect to the growing use of wireless computing in our society that is more subtle than any new technology. It's the next generation of users and their affinity for and rapid ability to adapt to new technology. Often labeled the generation effect, it has been observed by the military where young new recruits can operate the electronic technology faster than the older trained personnel. The effect may be more appropriately called the socialization of rapid adaptation to technology. Or, if you have kids like mine, you can call it the video game and instant messaging school of technology. Regardless, it is real, and some portion of workers today, and almost all of them tomorrow, will be very comfortable with small screens, tiny keyboards, lots of layered menus, micronavigators, and all of the changes that rapidly follow.

Wireless is moving into the mainstream of education and will propel the socialization of our next generation of customers and employees. As discussed in the chapter that included wireless in education, in 2001 the University of South Dakota became the first university in the United States to require handheld computers for first year students. University President James Abbot said, "The University of South Dakota students live in a mobile society. We must provide a learning environment using the latest technology so our students can take advantage of the benefits of anytime, anywhere learning to better prepare for the future."[6] Visit Consolidated High School in Orland Park, Illinois, just outside of Chicago, where 1,700 students and 65 teachers all use Palm Pilots. Need some more examples? Palm awarded grants to 80 schools selected from more than 1,200 schools that applied for Palm support. Microsoft is providing pocket PCs to seventh and eighth graders in Eminence, Kentucky. The curriculum materials are translated into Spanish and provided to migrant students on their pocket PCs. Last, Texas Instruments formed a partnership with goReader to offer educators a pen-based tablet with a 10.4-inch color high-resolution screen that holds more than a year's worth of textbooks and allows note taking and bookmaking. Do you remember when Apple seeded PCs in elementary schools?

Kids aren't waiting to get wireless devices from school. Not surprisingly, 4 out of 10 children in the United States between the ages of 4 and 18 owned some kind of wireless device in 2002. The study was done by Circle 1 Network and SpectraCom and showed a marked increase from 2001. Do you believe the next generation will fully adapt and become comfortable with handheld wireless computing devices? Count on it!

In 2000, I led an acquisition of a small company in Scotland. We were very impressed with this company because it had gone totally paperless in a very paper-intensive industry. As I toured the office, I saw that each worker had a scanner on his or her desk. Paper sent in by clients would immediately be scanned and then thrown away. By going paperless, the company was able to replace hundreds of square feet of cabinets with its document management system. The company e-mailed everything to the clients and received most of everything back in e-mail form. During a visit to this company's office, I had some documents that I needed to fax back to my headquarters. I searched all around the office looking for a fax machine and couldn't find one. Finally, I asked for help, and a buyer showed me the one outbound fax in the office, hidden in a corner. I asked him to show me the appropriate keystrokes for sending, and he suddenly got a puzzled look in his eyes. "I don't know how to send a document on this," he said sheepishly. I had just witnessed a microcosm of technology evolution and adoption.

As wireless moves from novelty to norm in our society, workers will become more and more comfortable with computing devices, various forms of data entry, and remote access to enterprise information. Following closely behind the employees are customers, who will come to expect their supplier's salespeople, mobile customer service reps, and service technicians to provide real-time information about products, their contract pricing, inventory availability, order status, specification sheets, and shipping details. The Internet has established a new standard for accessibility and immediacy of information. Customers are getting used to the convenience and control possible with an always-connected supply base over the Internet. It is logical that the expectation of always-available information will eventually be extended to wireless. Bill Gates believes it and has said that he believes 50% of all Internet traffic will come from wireless by 2003. He may be a little early on the date, but the trend is clear.

REFLECTING ON THE CHAPTER

I hope this chapter has helped you get more comfortable with the premise that in the near future we will be quickly moving toward the mainstream use of wireless for practical business applications based on the three macro forces of change. The mobile workforce is growing, and workers need to be connected to the enterprise to improve productivity. The competitive drivers for business are compelling. The barriers to widespread business adoption of wireless are fading as network infrastructure and device performance continue to take a huge leap forward in performance. On the horizon there is a massive wave of new mobile wireless Internet users with expectations of real-time mobile connectedness in every aspect of their lives.

We are at a different place today than just a few years ago when wireless pioneers like FedEx invested what were considered staggering amounts of money to achieve a sustainable competitive advantage. What's so exciting is that the Office Depot system previously discussed was deployed for roughly 10% of its equivalent system cost per user compared to the original FedEx system. There are certainly scale issues to consider, but the point is that like general computing, today's wireless costs for an enterprise are lower and carry less risk than in the past.

All of the factors and forces of technology, culture, and costs are converging to push a change wave that is driving business wireless forward. Wireless for business is going mainstream. What is your company's strategy for wireless?

ENDNOTES

1. "Economist Intelligence Unit and Infosys Technologies: Uncovering the Real Value of Mobile Computing," *ebusinessforum.com* (November 1, 2001), access at *www.ebusinessforum.com.*

2. Mike Drummond, "Wireless at Work," *Business 2.0* (March 6, 2001).

3. "No Stopping WLANs," *PC Magazine* (November 19, 2002): 27.

4. Margie Semil, "Wireless Takes to the Trenches," *M-BusinessDaily* (June 28, 2001), access at *www.mbusinessdaily.com.*

5. See note 1.

10

Looking Ahead

More technological innovations have failed than succeeded, by a factor of 100 or more. Just in the last 10 years, billions of dollars have been spent on what were perceived to be the most promising ideas. Unfortunately, the practical aspects of customer preferences or behaviors were glossed over by the force of a hopeful vision.

As business leaders, it is critical to embrace the potential of new ideas while seeing through the hype and euphoric vision statements to a practical view of the future. Although this book is grounded in the practical, it is also intended to promote consideration of what might be possible. The future often arrives earlier than we realize possible especially with technology. I've identified what I see as the key breakthrough innovations and applications that will have a real impact on commerce in the near term. These are ideas that you probably have heard about, but that are now quickly moving from the possible to the practical with promising pilots completed and more applications in development.

WIRELESS INTERNET WILL BE EVERYWHERE AND IN EVERYTHING

Although I'm always skeptical, the more I hear this, the more it makes sense. Think about where the Internet is available today—more places than you would think. You can access it at the ATM, in coffee shops, at airports, hotel rooms, school rooms, bedrooms, in planes, and through your phone.

The Internet is the universal platform for networking and communications among people, their machines, and their environment. Several infrastructure and device technologies are taking all of us toward the anywhere, everywhere, anytime Internet destination. Just around the corner,

WLANs will be almost everywhere. The cumulative impact of the upgraded telecommunications networks, together with the rapid growth in WLANs, add up to a connectivity critical mass. Given that reality, it isn't too much to expect wireless to extend the Internet to everywhere else, including cars, subways, trains, and sidewalks, and in everything, including vending machines, toll booths, security systems, ATMs, appliances, and medical monitoring devices.

The concept of the personal area networks (PAN) describes an environment that connects me and my devices, via a wireless link, to one of millions of Internet network nodes. Wherever I go, I will be on the Internet, if I wish. Even more so than today, work will not be someplace you go—it is something you do where you are.

Just as the Internet will be everywhere, the capability to communicate with the Internet will be in almost everything. It's closer to reality than you think. Of the 5 billion microprocessors sold in the year 2000, only 120 million (2.5%) were built into PCs. The largest amounts of computing power were embedded into manufactured products. It's called pervasive embedded machine intelligence (PEMI). Just look inside your new car! If it is a BMW, it has more than 100 embedded chips. In addition, the typical middle-class home has more than 40 microcontrollers included in things like electronic scales, self-regulating irons, and security systems. According to Harbor Research, by 2005, the estimated number of chips in the average home could reach near 300.[1] By 2007 the annual rate of embedded chip production (non-PC) may reach more than 9 billion chips.

The growth in PEMI chip production evidences the trend for electronic monitoring, sensing, and surveillance in our society. There will be growth in sensors for traffic flow, chemical agents, occupancy, and global positioning. Smart buildings will use the sensors to automatically operate heating, venting, and air conditioning, security, lighting, and sound masking. Building maintenance personnel will be connected to a building's pulse and will be instantly alerted to level, pressure, and temperature alarms. Personnel will also be able to remotely access and control pumps, motors, cooling systems, and lighting controls. Additionally, security for buildings and public places will become more prevalent with increasingly sophisticated video sensors to wirelessly notify the security company and its roving guards of any problems. PEMI will permeate our health care. Telemedicine, the process of diagnosing and/or treating patients remotely (discussed more later), will extend to patients wearing devices that monitor temperature, heart rate, respiration, sugar levels, and one day, even brain waves.

Eventually, intelligence will be embedded into the most common objects and materials with which we interact. Remember radio frequency identification devices (RFIDs)? RFIDs will keep getting better and smaller. RFID is taking a giant leap forward in utilization. Forty of the biggest companies in the United States, including Wal-Mart, are participating in a Massachusetts Institute of Technology project to standardize RFID systems. Standardization will allow products to flow throughout the supply chain and still be tracked. Store shelves will be equipped with RFID readers that will sense the RFID on every bottle of shampoo and will be able to take their own inventory—all the time. More accurate inventory management will improve pull signals from manufacturers. Shipments will be more accurate as well. Just as the UPC code revolutionized the ability to identify consumer products (what it is and how much to charge), RFID will add the attributes of location and visibility.

The impact of embedded RFID extends beyond manufacturing, through distribution, to consumption points, and then ultimately to disposal. RFID devices will follow a version of Moore's law by decreasing in size and cost rapidly over the next few years. How far will it go? International Paper has developed technology embedding an RFID chip in the cardboard container of milk. In the near future you will be able to walk through the grocery store with your PDA, select and compare items for purchase, and then add them to both your virtual and your literal shopping cart. There will be no checkout lines for the wireless empowered.

Until recently, we have made slow progress networking devices together; however, the anywhere Internet, together with the upgraded wireless telecommunications infrastructure, is allowing that networking. Harbor Research predicts that by 2010 more than 500 million devices will be Internet connected. There are more than a few hurdles to overcome to figure out how to get all of the chips and devices to talk to each other. Just as XML provided a common frame of reference to facilitate Internet-based communications, similar standardized protocols or alternative gateways will emerge with wireless.

Imagine, if you will, a digital nervous system connected to everything around us. Smart buildings, smart cars, smart health monitors, and smart power. You will use your personal communications tool to access and connect a multitude of objects and machines to your PAN. We may even end up with smarter people. How will your company connect to the digital nervous system?

STREAMING MEDIA

Streaming media (live video) to a wireless device is discussed as one of the potential benefits of the 3G high-speed networks. This means you can watch a football game or an emergency address by the President over your wireless device, while waiting at an airport or walking down the street. Entertainment is powerful, but the real business application for streaming media is video-teleconferencing. In business, the value of seeing the face of your customer or co-worker is high, possibly high enough to be worth the cost; however, although most salespeople would want to have more pseudo-face time with customers as opposed to voice only, they would resist substituting virtual face time for real face time.

Many are enamored with the idea of streaming media, but I believe that streaming media will be slow to catch on mainly because of the price/practicality equation. What hasn't quite sunk in yet is the recognition that moving full-motion video over a network transmission link to a wireless phone will be an expensive call. On the wireless user end, it will consume lots of data bytes and a good bit of battery time. It will be expensive to deliver and expensive to consume. There are alternatives being developed, such as using millions of lower-cost WLAN access points (802.11b or a) that are hardwired to the Internet. Using a WLAN provides wireless devices a short hop to a lower-cost high-speed network. In order for this to really work, WLAN access points will have to be almost everywhere, especially in malls, airports, office buildings, coofee shops, and every gas station. It is already happening as you read this.

Several business models are working toward a nationwide pervasive WLAN access strategy (Boingo and Wayport being two pioneers), but for the general population even in the major cities, the cost of deploying the infrastructure for a pervasive WLAN network is significant. The companies that are investing in the expensive high-speed 3G networks do not like the aforementioned scenario for obvious reasons. The good news is that with a converged device (both telecommunications and WLAN capability, discussed more later), benefits can be derived while the WLAN grows. Meanwhile, telecommunications network costs should drop.

Companies that are investing in 3G will only get a good return if a large block of users adopt and use lots of bytes of data. We learned from the failed Iridium satellite model that cost to the consumer does matter. To attract consumers in large numbers, audio and video applications will have to have compelling content, widespread distribution, and reasonable pricing, and even then it will be a tough sell because the average user in the United States views the wireless phone primarily as a voice tool, rather than a data communications or content delivery tool. Look at it another way: Why would you want to view a movie, a game, or the President's speech over your handheld at a premium price? Why wouldn't you watch it on your big-screen digital TV, or if you are on the road, on the many TVs in hotel lobbies, bars, or in your room for free? I'm not saying that handheld streaming media does not have some appeal; it just doesn't seem to have compelling value, with the exception of mobile video-teleconferencing. Over time prices must come down for video streaming to have widespread adoption.

REAL-TIME ECONOMY

Pervasive computing over the Internet will increase demand for real-time information. Compare for a moment your tolerance for waiting for information (or anything) from 10 years ago to today. Before FedEx, it didn't absolutely have to be there overnight because it just couldn't get there overnight. Now, we use FedEx, UPS, and DHL as a supplemental mail service, and accordingly a huge amount of documents have moved to overnight delivery as a standard. In the early 1990s, e-mail hit the mainstream and the pace went even faster. Network speeds have increased and wireless speeds will increase soon as well. Everything is speeding up approaching real time. Companies or processes that do not move toward real time will be negatively compared to and stand in the shadows of those that do. Our culture may be the only one in which people stand in front of a microwave oven tapping for it to hurry up. Nobody wants to wait for anything anymore, especially information. How real time are your business communications?

HANDHELD DEVICE OF THE FUTURE

Wireless devices are undergoing rapid and dramatic changes. Their processing power, memory, display, and communications capabilities improve significantly continuously. The primary innovations of the future will be larger, brighter, clearer displays, more processing power and memory, and dual-mode communications capabilities.

The premier wireless device of the future will hold more than 10 gigabytes of information, run at speeds of 1Ghz or more, have a super-bright display, dock and synchronize with network systems, and communicate equally well and seamlessly with public WLANs and both 2.5G and 3G telephony standards (GSM and CDMA). In addition, the device will have GPS capability and an extended battery life that is four to six times longer than current batteries. All this and it will be under $1,000.

Then there is convergence. The term *convergence* is used to describe the predicted blending of three key technology elements into a single powerful wireless computing and communications device. Imagine a device that had some of the capabilities of a PC, the functionality of a PDA, the usability of a phone, and multimode connectivity capability allowing it to connect to the Internet via one or more high-speed telecommunications network paths or an 802.11b(or a) WLAN. All of the most desirable and necessary functions would be combined efficiently and economically into one device or family of devices. Convergence is on the immediate horizon.

There will also be innovation with specialized types of wearable wireless devices, some personalized for specific kinds of jobs, others for the general consumer. A Gartner study predicts that from 2006 to 2011, always-on wearable wireless will become mainstream.[2] Gartner boldly forecasts that 60% of the population ages 15 to 50 will wear a wireless communications or computing device at least six hours a day in 2007. Continuous upgrading of performance while shrinking their size, together with connectivity enhancements, may make the prediction a reality.

On the business side, large companies like FedEx and UPS will have a significant influence on device designs in their unique vertical markets. Families of similar devices will form around vertical market segments of health care, logistics, service, and transportation.

The devices of the future will be always connected to the Internet, making synchronization real time. Smart phones and especially wireless tablet PCs will continue to grow in popularity, but will laptops ever go away? Maybe, but only when the handhelds are fully compatible with corporate business system requirements and applications and when the best features and functionality are available in one converged device.

VOICE RECOGNITION WILL TRANSCEND THE FORM-FACTOR PROBLEM

The final threshold to unfettered usability for a wireless device is the method of entering data. The device of the future will allow users to speak their requests and commands directly into their phone to search, sort, order, and e-mail at will, as well as enter keystrokes via a stylus or keypad. Voice recognition technology, when fully developed, will forever mitigate (but not completely eliminate) the form-factor problem of data input on a wireless device. Interactive voice response systems will improve dramatically, allowing voice-only access for more and more functions. Voice browsing will become a convenient reality. Voice recognition and supporting technology will get really good, really fast.

Voice recognition improvements will also fit nicely into telematics—wireless computing within automobiles. In addition to features recently available (see Chapter 4, Wireless for Customers), onboard computing will improve dramatically and enhance existing navigation, fuel stop and restaurant identification, text-to-voice translation of e-mail, both receiving and sending, and emergency assistance. All of these tasks can be done hands-free in the car with minimal distraction; however, going to the next level of Web access for surfing or to transactions for the driver will be more challenging and may be blocked by regulatory initiatives aimed at limiting driver distraction. That doesn't mean that the passengers can't enjoy satellite-fed movies, Web-based gaming, and general Internet access. After some fine-tuning and testing of features, lowering of costs, and broader user acceptance, the automobile will be an important computing center.

Finally, voice recognition systems will work together with Softbots to make wireless phones ever more productive. Softbots are software programs that are customized to accomplish specific searching tasks on the Internet. For example, one would pick up a wireless phone, hit a hot-key, and speak "Locate a Mexican restaurant, medium price, on the south side of town, two choices with ratings." The Softbots would then run the search while you drove or finished a meeting, and then would push an alert to you to pull down the information. No searching or surfing. This could be fun.

VOICE OVER THE WEB

Transmitting voice over the Web is becoming commercially viable based on improvements in technology and upgrading of network quality. Working closely with voice over Internet protocol is the session-initiated protocol (SIP) phone. SIP phones are really software programs resident on PCs. The user speaks the name of the person he or she wishes to call, and the request is sent to a voice-activated dialing application and then to the SIP server, which routes the call over the Internet to the requested recipient's ordinary phone. This makes great sense because companies have invested a lot in their data networks, so why not let it do double duty and route voice calls along with the data?

BIOMETRICS

Biometrics allows a computer to confirm an individual's identity using a stable physical trait or biological signature, such as a retinal scan, fingerprint, voice analysis, facial recognition, dynamic signature analysis, and some time soon even DNA. Biometrics is the most promising technology to overcome the inherent weaknesses associated with username and password because it verifies the human traits that cannot be passed to unauthorized users. For example, Fujitsu is selling fingerprint sensors for its pricier wireless phones. The sensors will make it more difficult, if not impossible, to use the voice or m-commerce capabilities of a stolen phone. Governments are the leaders in utilizing biometric technology. Instead of logging onto their networks with users' names and the typical easy-to-crack passwords, about 2,000 city employees in Glendale, California, now use their fingerprints to sign on to their PCs.

Biometrics is becoming big business. DataQuest estimates the market for biometric technology to be $2.6 billion by 2004 from almost zero in 1998.[3] An association has been created to promote standards and use of the technology, the International Biometric Industry Association. The big plus for consumers, besides the enhanced security, is that you don't have to remember anything!

BEYOND TELEMEDICINE

Wireless is already permeating the health care system. In the future, wireless will be used to advance care even further. Patients will wear life bracelets, which constantly transmit their vital signs, and in the case of elderly patients, their location and physical position (in bed, walking, etc). The advanced monitoring technology will also allow better management of acute illnesses because wireless monitoring devices will be attached to diabetes patients to monitor their glucose levels all day, specifically after taking drugs undergoing field trials for approval. Wireless health bracelets will also contain your entire medical history, blood type, allergies, and current medications for paramedics to scan after they locate you in an accident.

In the midst of all of the wireless innovation in the works, practicality and utility have always and will continue to prevail. That doesn't mean there won't be plenty of wireless gadgets to distract us from those that add real value. In the category of potential wasted billions, take the wireless beer glass. It emits a radio signal when it is near empty, alerting wait staff (on their PDAs of course) of the need for a refill. Will this be the service breakthrough we have all been hoping for?

REFLECTING ON THE CHAPTER

The future for wireless applications is powerful and almost overwhelming in its scope. I recall the phrase I often heard some years ago: "Wireless will change everything." Together with that powerful statement was an implied urgency that wireless *would* change everything—right now! I didn't really believe it then, but I do now, with a qualifier. Wireless will change *many* things, probably more than we can imagine, and it will happen one practical application at a time.

ENDNOTES

1. "Pervasive Internet Report," Harbor Research (April 2002), access at *http://harborresearch.com/pir_demo.*
2. Matt Hamblen, "Grin and Wear It," Synchrologic (April 26, 2002), access at *www.synchrologic.com.*
3. Jason Wright, "Biometrics: Is it Making a Splash Yet?" *SC Magazine*, 25.

Appendix A

First Practice
50 Wireless Enterprises

COMPANY	INDUSTRY	FOCUS
1. Ace Beverage	Distribution	Beverage
2. Associated Food Stores	Distribution	Grocery
3. Avera McKennan Home and Community Services	Health care	Home health care
4. Avecra Oy (Finland)	Food services	Passenger rail service
5. Averitt Express	Transportation	LTL freight carrier
6. Bartlett Tree Experts	Tree care	Sales and service
7. Beth Israel Deaconess Medical Center	Health care	Patient care
8. Boston College	Education	University
9. Boston Public Schools	Education	Truancy control
10. British Petroleum	Petrochemical	Maintenance
11. Carlson Hotels Worldwide	Hospitality	Hotels
12. Celanese Chemicals	Manufacturing	Chemicals
13. Cemex	Manufacturing	Construction materials
14. City of Fairfax, Virginia	Municipal government	Public transportation
15. City of Richmond, British Columbia	Municipal government	Flood monitoring
16. Coors Field	Entertainment	Concession sales

17. CSX Corp.	Transportation	Rail
18. Daytona Beach Police Department	Municipal government	Police department
19. Dedham Medical Associates	Health care	Billing/accounts receivable
20. Delta Employee Credit Union	Financial services	Consumer banking
21. Dunkin' Donuts	Food services	Retail fast food
22. Equitable Insurance, Inc.	Insurance	Life insurance
23. Famous Footwear	Retail	Footwear
24. Fidelity	Financial services	Consumer brokerage
25. First Service Network	Service	Building maintenance
26. Florida Power & Light	Utility	Metering
27. GE Medical Systems	Service and repair	Medical equipment
28. Harris Bank	Financial services	Retail banking
29. Illinois State Police	Government	Law enforcement
30. Interbrew (Belgium)	Food service	Beverage
31. Intercoastal Realty	Real estate	High-end home sales
32. Jim Hudson Lexus/Saab	Automotive	Retail
33. JS Express	Transportation	Courier
34. McKesson Corp.	Wholesale distribution	Pharmaceuticals
35. McLane Co.	Distribution	Grocery
36. Mesa Energy Systems	Service	HVAC
37. Miami Dade Water and Sewer	Municipal government	Water and sewer
38. Moses Cone Healthcare System	Health care	Hospital patient records
39. Owens & Minor, Inc.	Distribution	Health care

40. Packaged Ice, Inc.	Manufacturing	Retail sales
41. Pfizer, Inc.	Pharmaceuticals	Manufacturing and shipping
42. PRI Automation	Manufacturing	Semiconductor equipment
43. Producers Insurance Co.	Insurance	Crop insurance
44. St. Luke's Hospital	Health care	Patient care
45. Scandinavian Airlines System	Airlines	Passenger travel
46. Starbucks	Retail	Beverage
47. Trinity Development	Construction	Highway
48. Vail Resorts	Recreation	Skiing
49. Visiting Nurses Association	Health care	Home nursing
50. Volkswagen (Germany)	Manufacturing	Auto

Appendix B

Directory of
First Practice Examples

This appendix provides brief profiles of the wireless initiatives of 50 companies and organizations across multiple industries. These pioneers represent the first wave of practical wireless utilization. Using a combination of custom applications and off-the-shelf tools, these leaders improved business processes, enhanced customer service, and solved business problems —all impacting the bottom line. We can learn from their accomplishments as well as appreciate their initiative.

COMPANY: Ace Beverage Co.
INDUSTRY: Distribution
FOCUS: Beverage

BUSINESS GOAL

Automate field force and enhance customer relationship management. Eliminate paper in route delivery processes and improve warehouse efficiency.

BACKGROUND/OPERATIONAL DESCRIPTION/RESULTS

Ace Beverage in Los Angeles delivers hundreds of thousands of cases of beer each year to chain and convenience stores in the greater Los Angeles area. Sales reps would visit each customer location to take their orders and share specials. The paper orders were brought back to the regional warehouses in the late afternoon. The orders had to be keyed in to the company enterprise resource planning (ERP) system before the process of sorting loads, trucks, and drivers began, which took more than an hour. Occasionally, salespeople would return late in the day, delaying the next day's loading and customer deliveries. Imagine dozens of sales reps all arriving late in the day with paper orders for computer entry, and then a mad rush through the night to pick, stage, and load trucks, which need to be rolling out the door by 5 A.M. the next morning.

With the new wireless system, salespeople can enter orders on a wireless tablet PC while at the customer location. Orders are then transmitted back to the central system, beginning at about 1 P.M. rather than arriving at the warehouse at 3 P.M. under the previous paper system. Now, orders for the next day are in to the warehouse by 3 P.M., rather than 8 P.M., enabling workers to start and complete loading the next day's orders much earlier, according to Mike Krohn, VP of Finance and Administration. Krohn says the two areas of savings are overtime and warehouse equipment costs. Because trucks are loaded earlier in the day, the company has been able to scale back its so-called graveyard shift, which used to begin at 8 P.M., and eliminate the need for two forklifts. Ace is also saving between 15 and 20 hours in driver and loading overtime costs each week.

Finally, salespeople can complete electronic order fasters than paper forms. That translates to more face time with customers on each call.

TECHNOLOGY

Sales reps use the Intermec Technologies Corp. 6642 Windows-based mobile pen tablets equipped with Sierra Wireless Air Cards. Wireless transmission is over an AT&T VPN. MiT Systems' EzSales for field force automation was used to replace the paper route pads with electronic forms and to gather the data and feed the data to the ACE IBM AS/400 system.

Source: *Pen Computing* (November 2001): 35.

COMPANY: Associated Food Stores
INDUSTRY: Distribution
FOCUS: Grocery

BUSINESS GOAL
Optimal management of delivery truck fleet.

BACKGROUND/OPERATIONAL DESCRIPTION/RESULTS
Associated Food Stores (AFS), a Salt Lake City, Utah–based food distributor, delivers to 600 stores in the Rocky Mountain area, shipping roughly $2 million worth of goods every day. Associated wanted to improve its management of a fleet of 70 tractor rigs and 400 trailers and lower fleet operating costs plus move inventory faster. Associated committed to a new wireless system to achieve its goals.

Using radio frequency indentification (RFID) tags on every trailer, and tracking software integrated with its fleet management system, Associated can now locate and track shipping containers in real time from its 600-acre distribution facility in Salt Lake City. Data are provided to dispatchers about each trailer, its location, and load status. For perishables, special tags monitoring temperature and fuel in the truck were added. The information is used by dispatchers to schedule a trailer and truck for its next shipment.

"WhereNet's technology allows us to track, locate, and monitor transportation assets and the contents of every load of lettuce, Popsicles, and chicken in real time," says Tim Van de Merwe, Internal Logistics Manager for Associated. Tim also says the company has been able to reduce the number of truck drivers from 100 to 62 and eliminate the use of 35 of 40 leased trailers, which usually handled the unscheduled overflows. More than 150 workers were previously involved in data entry, walking around the yard, noting the whereabouts of trucks, locating special trailers, checking loads, tracking shipments, and scheduling deliveries. With the new software and technology, it takes only two people to manage the system.

TECHNOLOGY
Associated Foods uses RFID technology from WhereNet Corp., Santa Clara, California.

Source: *Information Week* (Dec 10, 2001): 24.

COMPANY: Avera McKennan Home and Community Services
INDUSTRY: Health care
FOCUS: Home health care

BUSINESS GOAL

Automate the entry and retrieval of patient care information and billing processes and improve the efficiency of mobile health care providers.

BACKGROUND/OPERATIONAL DESCRIPTION/RESULTS

Avera McKennan Home and Community Services of Sioux Falls, South Dakota,
provides 24-hour care to patients in their homes as an alternative to higher-cost
stays in hospitals. In 1995 Avera developed a software package to help automate
recordkeeping and reporting.

Avera needed a mobile computing solution that would allow doctors and
nurses to avoid before- and after-shift trips to the office to gather patient records
and drop off notes for manual updating the following day. Avera successfully
extended the software to handhelds, so doctors and nurses could retrieve records,
complete forms and access care plan databases from the road. Caregivers in the
field now download their daily assignments, complete forms while in a patient's
home, and upload the information for streamlined processing. No trips to the
office are required.

Caregivers can now fit in one more patient visit per day, generating more
revenue and saving time and mileage costs. Double entry of patient information
has also been eliminated, saving thousands of hours of staff time. More important,
recordkeeping is more precise and less prone to errors caused by transcription or
double entry.

TECHNOLOGY

Microsoft Windows CE powers NEC Mobile Pro devices provided by Patient
Care Technologies in Atlanta.

Source: Microsoft case studies: Avera McKennan Home and Community Services
(*www.microsoft.com/resources/casestudies/company.asp*).

COMPANY: Avecra Oy (Finland)
INDUSTRY: Food services
FOCUS: Passenger rail service

BUSINESS GOAL
Improve control over sales and material costs. Respond to inventory needs faster.

BACKGROUND/OPERATIONAL DESCRIPTION/RESULTS
Avecra Oy, the catering services provider for Finland's national railway, uses a wireless system to transmit real-time sales and inventory data between railroad dining cars in transit and its central computer system. Waiters are equipped with handheld PDAs that continually search for connection points to the Internet as trains traverse their routes. Once a connection is made via the 802.11b wireless network, real-time inventory and sales data are securely transmitted to Avecra's headquarters in Helsinki. Based on the steady stream of data incoming from railway cars, Avecra can replenish quick-selling menu items at intermediate station stops.

In 2001, Matti Saari, Avecra's financial manager, said the company was the only one he knew of that used such a wireless system onboard train cars. Avecra estimated it would take between two and five years to see a positive return on its investment of $180,610. In the meantime, the catering services company has gotten much better control over its inventory and sales records and efficiency, Saari said.

TECHNOLOGY
NetSeal Technologies' RoamMate software, Espoo, Finland

Source: Alan Radding, "Leading the Way on Wireless," *ComputerworldROI* (September 2001).

COMPANY: Averitt Express
INDUSTRY: Transportation
FOCUS: LTL freight carrier

BUSINESS GOAL

Automate logistics and shipping records to reduce costs and delays associated with errors and misrouted shipments.

BACKGROUND/OPERATIONAL DESCRIPTION/RESULTS

Cookeville, Tennessee–based Averitt Express is a transportation company that delivers freight from Virginia to Florida to Texas via 92 regional facilities throughout the Southeast. Averitt moves approximately 17,000 shipments per day representing 20 million pounds of freight. The process of tracking cargo, determining destinations, and scheduling was done manually with a log system and lots of paper. As a less-than-truckload (LTL) carrier, Averitt would pick up loads from various cities, consolidate them in its regional centers, and ship them to their final destination. With the paper system, 3.5% of shipments were misrouted, costing time and causing customer frustration. Averitt was using an AS400-based system with customer logistics software to manage the complex combinations and routings; however, the company had no automated method to enter receiving and shipping information. Averitt decided to deploy wireless handhelds to streamline the process.

Dock supervisors began using wireless handhelds to enter real-time information directly into the AS400 system via a WLAN. Operators can receive directions on which dock to take certain shipments without needing to first locate paperwork. Errors have been reduced to one-half of 1% and visibility of shipments is accurate and real time, according to Will Taylor, senior engineer. Dock workers are also more productive because movement instructions are pushed directly to their handhelds, rather than to their foreman, who then used to relay the information to them. Efficiency has improved, with total freight throughput increasing by 3%; more goods on fewer trailers makes more money for Averitt.

TECHNOLOGY

Averitt uses IBM's AS400 operating system and Skaneateles, New York–based HHP's Dolphin handhelds, which communicate directly with the AS400 system (no middleware).

Source: Becky Swissler, "Dolphins on the Dock," *Field Force Automation* (August 2001).

COMPANY: Bartlett Tree Experts
INDUSTRY: Tree care
FOCUS: Sales and service

BUSINESS GOAL

Improve productivity of sales process. Streamline proposal generation process. Improve quality of customer interaction and customer service.

BACKGROUND/OPERATIONAL DESCRIPTION/RESULTS

Bartlett's 180 arborists are located in North America and Europe. They advise clients about tree care and pruning services. After visiting and assessing a customer site, they used to return to the office to prepare a proposal, then fax or mail it to the customer. A third option was to make an additional trip back to the customer site and personally deliver the proposal. The company's 89 offices had databases that held customer information, but the database was not accessible by the field salespeople. Instead, salespeople carried reference books to help diagnose tree diseases and develop treatment plans. Time delays, slow responses to customers, and a lot of paper convinced Bartlett that the salesforce was ready for some technology.

Using new tablet computers with special graphics and mapping capability, salespeople can now graphically map a customer's site, identifying and describing each tree. They can also take digital pictures of the trees and store them for later retrieval and future comparison. The salespeople can also access a diagnostic database, which contains tree health care reports and the latest information. Most important, salespeople at client sites can generate customer proposals—on-the-spot, answering questions, and closing the order. For subsequent calls, the salespeople can retrieve all of the information and be well informed in front of the customer.

The number of customers has almost doubled since the system was deployed in early 2001, while the number of salespeople has held steady. The initial cost of the system was near $1 million, representing one of the largest investments in technology this $120 million company has ever made. A positive return on investment, generated largely by a reduction in paper processing and travel time, is estimated at four years. This, however, does not include the productivity increases for salespeople, who have increased coverage to additional clients.

TECHNOLOGY

Fujitsu Stylistic LT C500 mobile pen tablet computers capable of clearly presenting color graphics and outdoor viewing. Software developed by BlueFlame of Hackensack, New Jersey.

Source: Mila D'Antonio, "BlueFlame Gets Tree Care Company Rooted in Mobile Sales," *CRM Magazine* (September 2001).

COMPANY: Beth Israel Deaconess Medical Center
INDUSTRY: Health care
FOCUS: Patient care

BUSINESS GOAL

Improve efficiency, reduce errors by optimizing patient record retrieval and updating and transmission and fulfillment of prescriptions.

BACKGROUND/OPERATIONAL DESCRIPTION/RESULTS

Beth Israel Deaconess Medical Center in Boston, Massachusetts, handles more than 60,000 patients annually and provides some patient wards with wireless Internet access and laptops to surf the Web. More important for operations, the hospital extended wireless to its emergency room check-in process. In the past, patients in pain would have to drag themselves to the desk and provide check-in information to a clerk. Today at Beth Israel, the clerks come to the patients' beds and gather the information on a wireless tablet. Patient information is updated on a new electronic dashboard in the ER command center for doctors and nurses to review. Replacing the old whiteboard and scribbles, the electronic dashboard displays key patient data in real time, along with lab values and other test results. As doctors examine patients, they input information into their wireless laptops, which update the hospital system and dashboard.

TECHNOLOGY

Dell Computer Corp. laptops with Cisco Systems, Inc., Aironet WLAN adapters

Sources: Danielle Dunne,"Wireless That Works," *CIO* (February 15, 2002); and Drew Robb,"Case Study: Care Group Health System," *Computerworld* (December 16, 2002).

COMPANY: Boston College
INDUSTRY: Education
FOCUS: University

BUSINESS GOAL
Wireless network access throughout a large multibuilding campus environment

BACKGROUND/OPERATIONAL DESCRIPTION/RESULTS
Providing network access throughout the dozens of historical buildings on the 116-acre suburban Boston campus was the challenge for Henry Perry, director of network services, and network engineer Brian David. Two factors convinced them to deploy wireless: cost savings and positive aesthetics. Hardwiring the buildings would have been much more expensive and unsightly, because it would involve ripping out ceilings and stringing cable in difficult-to-access locations throughout the historic buildings.

Faculty and students can now use laptops equipped with wireless LAN cards to access the school network from more than 350 wireless access points throughout the campus. These access points are located in dorms, lobbies, cafés, the library, and outdoor quadrangles. Students can register for classes, view curricula, and use e-mail from wherever they may be on the campus.

TECHNOLOGY
Enterasys Networks based in New Hampshire

Source: "The Wireless 25," *ComputerworldROI* (September 2001).

COMPANY: Boston Public Schools

INDUSTRY: Education

FOCUS: Truancy control

BUSINESS GOAL

Provide truancy officers with a more efficient means of accessing student records and related data during encounters with students throughout the city.

BACKGROUND/OPERATIONAL DESCRIPTION/RESULTS

Boston Public Schools has 63,000 students. Most attend school regularly, but more than a few are truant, wandering the city during school hours. Boston Public's truancy officers spot likely truant students and ask them for identification. Before to the new wireless system was implemented in April 2002, the officer would then pull out a document the size of a city phone book and leaf through thousands of pages attempting to locate the student. Later, paper forms would be filled out and sent in to update the student's record.

Today the phone books and paper forms are gone. Truancy officers now carry Web-enabled phones that quickly access an online repository of student and legal records, such as warrants and arrest records. Truancy officers can also access and update student school records on the spot. The new phones have a two-way fast-connect feature so officers can link to each other quickly. In the future, the school district plans to add barcode readers to scan student ID cards, speeding information input. Truancy Officers can check more students per day and keep the records updated with less overall effort.

TECHNOLOGY

Reston, Virginia–based Nextel Communications' i85 Java-enabled phones with digital e-service operating with Blue Bell, Pennsylvania–based AirClic's Mobile Information Platform

Source: "Boston Public: Snagging Class Cutters Wirelessly," *Wireless Week*, online magazine (April 3, 2002). See also "Boston Schools Equip Truancy Teams with Wireless Data Access," *M-Business Daily*, access at *www.mbusinessdaily.com.*

COMPANY: British Petroleum
INDUSTRY: Petrochemical
FOCUS: Maintenance

BUSINESS GOAL

Provide mobile plant workers with wireless access to equipment history and performance information, improving productivity and enhancing the visibility of key equipment performance indicators.

BACKGROUND/OPERATIONAL DESCRIPTION/RESULTS

British Petroleum (BP), the worldwide producer of petroleum and petrochemical products, has numerous refineries around the world that produce millions of barrels of products every year. Each refinery site is hundreds of acres in size, with equipment and critical processes throughout.

One BP location in Texas, which is ranked as one of the top three largest in the United States, covers nearly two square miles. BP wanted to equip its maintenance workers with wireless units that gave them access to critical performance information on key pieces of equipment, plus enabled them to record instrument and environmental readings and update the refinery systems in real time. Workers collect and input pressure, vibration, temperature, safety, and environmental data on their rugged handheld units, which instantly update the refinery systems in real time. The system includes a maintenance program that predicts required maintenance and provides details on equipment history and parts required, all of which is available to maintenance workers anytime and anywhere throughout the plant.

By taking a proactive approach to equipment monitoring and maintenance, BP aims to extend equipment life, reducing downtime and unexpected shutdowns, all ultimately contributing to greater plant output of petroleum products, which means more revenue.

TECHNOLOGY

Symbol Technologies' PDT 8146 rugged handheld running on a Spectrum24 wireless network. Software is by Houston–based systems integrator Systems Application Engineering (SAE), *www.saesystems.com*.

Source: "BP's Worldwide Refineries to Roll Out Wireless Mobile Computers," *MobileVillage.com* (January 23, 2002).

COMPANY: Carlson Hotels Worldwide
INDUSTRY: Hospitality
FOCUS: Hotels

BUSINESS GOAL

Improved information flow between key occupancy, pricing, and reservations systems and roving hotel managers

BACKGROUND/OPERATIONAL DESCRIPTION/RESULTS

Minneapolis-based Carlson Hotels owns more than 765 hotels in 64 countries, including Radisson and Park Plaza Inns. Carlson invested more than $20 million revamping and integrating multiple systems and databases to create a streamlined back-office and reservations system, which includes critical information such as room status and occupancy levels. Until 2002, however, the system was limited to desktop access. As managers and supervisors walked the properties, they could not reach the valuable information.

Carlson deployed wireless handhelds to all managers linked to the hotels' management system. Managers can now monitor sales, operations, and key account information from anywhere on the hotel site. Using the new tools, managers can monitor and optimize room rates on a continual basis, receive urgent messages, and access the corporate systems, all while walking around their properties. Managers can also customize alerts and color code them based on urgency. The new tools have made Carlson managers more efficient and more in tune with their business. Instead of waiting for reports a week after they are needed, the handhelds can access the information anytime and from anywhere on the property.

TECHNOLOGY

Handhelds are the Compaq Ipaq 3630 and 3650 pocket PCs running Windows CE 3.0, synchronized with the corporate systems and databases via a CISCO 802.11 PCMCIA expansion slot LAN card communicating with a Microsoft Mobile Information Server.

Source: "Wireless That Works," *CIO* (February 15, 2002); and Mitch Wagner, "New Users, New Devices," *Internet Week* (May 4, 2001): 1.

COMPANY: Celanese Chemicals
INDUSTRY: Manufacturing
FOCUS: Chemicals

BUSINESS GOAL

Improve customer service and increase revenue by providing salespeople with anytime, anywhere access to product and pricing information.

BACKGROUND/OPERATIONAL DESCRIPTION/RESULTS

When customers would call salespeople from Dallas–based Celanese Chemicals Ltd. with questions about their accounts, they would politely excuse themselves from the call, then quickly call customer service personnel who had network access at the home office. After retrieving the required information on products or orders, customer service would call back the salespeople and relay the information. The salespeople would then call back the customer with an answer. The process took from four hours to a full day. If there were more questions, the process was repeated. Wireless changed all that.

Now, mobile salespeople use wireless handheld units linked to a Web server that accesses the company's SAP system and retrieves information in real time. Using the wireless system, salespeople can locate customer shipments, enter orders, and track delivery information to give to customers directly. Celanese expects to see an increase in customer satisfaction and subsequent improvements in customer retention and penetration.

Celanese receives more than 50,000 orders per year totaling $1.5 billion. Because of the size of the orders, it doesn't require saving many formerly missed orders to pay for the system. The total cost of the program was less than $50,000, with each wireless device costing between $500 and $1,000 to purchase and outfit with the software.

TECHNOLOGY

SAP/R3 is the ERP system. GadgetSpace provided the software that takes the information from SAP and reformats it for handheld devices.

Source: Matt Hanblen, "Taking the Leap," *ComputerworldROI* (September 2001).

COMPANY: Cemex

INDUSTRY: Manufacturing

FOCUS: Construction materials

BUSINESS GOAL

Optimize the delivery of cement, a perishable commodity, across multiple sites.

BACKGROUND/OPERATIONAL DESCRIPTION/RESULTS

Cemex in Mexico City is the world leader in cement manufacturing and delivery. The cement business is unique in that the cement, once mixed, is immediately perishable and must be poured within 90 minutes of mixing. If poured too late, the entire batch is lost, and any unused materials must be quickly poured at another client site or poured out on the ground and wasted. Cemex sought to improve the timing and routing of trucks to both avoid waste and optimize use of its delivery fleet. The challenge was to route the mixer trucks from the mixing plants to the building site as required by the client, then route the trucks to other sites, all within the narrow time limits, then return the trucks for more materials.

Cemex deployed a computer and global positioning receiver in every truck, combining the location information with output from various mixing plants and with the current orders from customers. The new software calculates which truck should go where and when with a specified amount of cement and includes dispatch instructions for redirecting trucks between deliveries. Even with the chaotic traffic conditions in Mexico City and last-minute rescheduling by customers, Cemex has achieved a sustainable increase in the number of orders filled per day.

TECHNOLOGY

GPS technology tracks trucks, which are linked wirelessly over a satellite-based communications network.

Source: Simone Kaplan,"Concrete Ideas," *CIO* (August 15, 2002). Accessed at *www.ebusinessforum.com* (18 June 2001).

COMPANY: City of Fairfax, Virginia
INDUSTRY: Municipal government
FOCUS: Public transportation

BUSINESS GOAL

Boost ridership on public transportation by enhancing customer service and
improving the monitoring of drivers.

BACKGROUND/OPERATIONAL DESCRIPTION/RESULTS

The Transportation Department of the City of Fairfax, Virginia, equipped 12 transit buses with GPS devices tied to an automated vehicle location system. The
result: real-time bus location and scheduling information delivered to passengers'
mobile devices. The information is viewable via Internet phones, PDAs, and on
digital display monitors at bus stops throughout the city. The system is designed
to delight riders by providing reliable scheduling information. In addition to
boosting ridership, city officials expect the new system to free up support employees
who used to respond to frequent calls for updated bus schedules.

The system also allows managers to monitor driver performance, recording
side trips, average cycle time, and time between stops.

TECHNOLOGY

System by NextBus Information Systems, Inc., Emeryville, California.

Source: "The Wireless 25," *ComputerworldROI* (September 24, 2001).

COMPANY: City of Richmond, British Columibia
INDUSTRY: Municipal government
FOCUS: Flood monitoring

BUSINESS GOAL

Automate the monitoring of 180 pumping stations while providing real-time status for faster response by field crews.

BACKGROUND/OPERATIONAL DESCRIPTION/RESULTS

The city of Richmond, in Vancouver, British Columbia, is surrounded by the Fraser River. The 160,000 citizens of Richmond are at risk when the river floods. They rely on a complex system of 180 pumping stations that work around the clock to keep the Fraser River from overflowing. In the past, city workers would monitor pumping stations by driving around and manually examining them—a process that was both time consuming and inefficient. Today, and for less than $100,000, the city of Richmond installed electronic monitors on each pumping station, which relay vital information to the city's central system and then on to field workers' handheld units—all in real time.

Workers can now respond to specific trouble spots immediately. The central control personnel can more efficiently deploy crews to the most urgent and critical problem areas. Wireless is helping keep Richmond above water. Since deploying the wireless flood-monitoring system in 2001, the city has also adopted wireless technology to track more than $1 million worth of inventory in its municipal warehouse.

TECHNOLOGY

New York–based Information Builders Inc. provided its Web Focus software for the Internet-data connection. Wireless providers are AvantGo Inc. in Hayward, Calif., and Sierra Wireless Inc. in Richmond, B.C.

Source: Danielle Dunne, "Wireless that Works," *CIO.com* (February 15, 2002); and 2001 *ComputerworldROI* survey of wireless innovators.

COMPANY: Coors Field
INDUSTRY: Entertainment
FOCUS: Concession sales

BUSINESS GOAL
Automate order entry, improving convenience for the customer and increasing sales.

BACKGROUND/OPERATIONAL DESCRIPTION/RESULTS
Coors Field is home to the Colorado Rockies baseball team in Denver, Colorado. Executives at the 50,000-seat stadium wanted to enhance the patron experience and allow fans to order concession items, such as hot dogs and pizza, from the comfort of their seats—no standing in line and missing a Rockies home run. To accomplish this goal, the roving food vendors use handhelds to take customer orders, scanning the row and seat numbers for accuracy before taking payment. The food orders are transmitted wirelessly via a WLAN for processing to the appropriate food station, where the order is prepared and then delivered to the customer's seat.

The handheld units deployed are made rugged for the environment, with simple icons for ease of use by food vendors. Customers are happy getting the items they want without the hassle of leaving their seat. Coors Field management now has access to previously unavailable point-of-sale information, such as food preferences by seating section. Vendors can also signal for additional help when fans are really hungry, something that didn't happen before the new system. The bottom line is increased sales of concession items, better information for future planning, and happy customers—a triple play.

TECHNOLOGY
Coors Field uses Microsoft Pocket PCs running software from Venue Tech, in Kalispell, Montana.

Source: Microsoft case studies: Coors Field (*www.microsoft.com/resources/casestudies/company.asp*).

COMPANY: CSX Corp.

INDUSTRY: Transportation

FOCUS: Rail

BUSINESS GOAL

Improve customer service by providing customers with real-time status on rail car location and related information.

BACKGROUND/OPERATIONAL DESCRIPTION/RESULTS

CSX Corp., based in Richmond, Virginia, is the largest freight railroad in the Eastern United States providing rail transportation and distribution services over a 23,000 route-mile network covering 23 states. Thousands of rail cars move throughout the CSX rail network every day. But previously, customers who wanted to know the status and locations of their shipments could obtain only very limited information. Now customers can track shipments 24/7 from Internet-enabled PDAs, using CSX's container search tool. Customers can track up to 20 rail cars at one time.

The application works similar to a Web-based locator application. In addition, customers can use cellular telephones equipped with wireless browsers to enter a rail car number and initial, and then receive the latest event information on that car. Additional information now on the CSX Web site will eventually be extended to the wireless system, including rail mileage calculators, schedules, average transit times, and reporting on weights.

TECHNOLOGY

CSX's proprietary wireless container search tool, which was developed inhouse by the company's CSX Technologies Group, works with any PDA device, including Palms, Pocket PCs, or other Windows CE devices.

Source: CSX Corporation Press Release (October, 17 2000), "CSX Transportation Leads Railroads Into Wireless Internet," *CIO* (March 15, 2001), CIO Special Report.

COMPANY: Daytona Beach Police Department
INDUSTRY: Municipal government
FOCUS: Police department

BUSINESS GOAL

Automate and streamline database searches for criminal records and outstanding warrants.

BACKGROUND/OPERATIONAL DESCRIPTION/RESULTS

The 130-member police department of Daytona Beach, Florida, is in control—at least until the population quadruples on Spring Break and during 10 other special events per year. During those times, the police force is stretched. Quickly obtaining information on detained suspects is key to keeping the force visible and available. Before the new system, officers would call in to a dispatcher and relay suspect information. The dispatcher would write it down and then access state and federal databases, eventually reporting the results back to the officer in the field. The officer and the detainee had to wait for the results. Besides the delays, officers were tying up radio frequencies with details of requests, often missing more urgent calls on other frequencies.

With the new system, officers use handhelds to access the criminal database directly and quickly, keeping radio traffic clear and saving time. Costs for hardware and software were approximately $50,000. The department estimates a payback of one year. Additionally, the system obviated the need to hire an additional dispatcher to handle information lookups, and tourists are detained for a shorter period, allowing officers to be more available and to handle more calls. For now, the wireless technology is CDPD, but eventually the city plans to upgrade to a higher-bandwidth network provided by AT&T Wireless.

Officers also now use the handhelds to file their arrest reports, which means spending less time at the police station filling out paperwork and more time in the field. Police also record information on each stop via their handhelds.

TECHNOLOGY

Software is from Burnaby, British Columbia, Canada–based Infowave Software, Inc., which converts information from the criminal databases so it can be accessed over the wireless cellular CDPD network on handhelds from Compaq Corp. and Hewlett-Packard Co. Visionair is the vendor of the department's dispatch and records management system software.

Source: Howard Baldwin, "2001 Mobile Master Awards for Enterprise Deployments," *M-Business* (January 2002).

COMPANY: Dedham Medical Associates

INDUSTRY: Health care

FOCUS: Billing/accounts receivable

BUSINESS GOAL
Improve accuracy and automate the patient diagnosis coding and billing processes.

BACKGROUND/OPERATIONAL DESCRIPTION/RESULTS
Dedham Medical Associates is a private practice outside of Boston with 75 physicians representing 12 different medical and surgical specialties. Dedham wanted to improve the efficiency and accuracy of its billing process. Bills are generated based on diagnostic codes that physicians record after treating each patient. Recording the correct codes and all applicable codes is key to accurate billing and reimbursement. Previously, physicians recorded the codes on a multipart paper form. The forms were turned in to the billing department, where they were entered into a computerized billing system. Errors in entry, transcriptions, and missed billing opportunities were costing Dedham money, as well as the time required to record the information twice.

With the new wireless system, physicians enter the complex codes on their Pocket PCs and then wirelessly transmit the information via an 802.11b LAN to the central billing system. With special software built to help physicians navigate complex coding rules, the wireless system has improved accuracy and streamlined the process, eliminating the double-entry system. The units also provide physicians with their current patient schedules and other information, helping them to be more productive.

TECHNOLOGY
Compaq Ipaq Pocket PCs running software named "Charges-in-hand" from MedAptus, Inc., Boston., on an 802.11b wireless LAN.

Source: "Dedham Medical Prescribes Handhelds to Its 75 Physicians," *www.MobileVillage.com* (January 23, 2002).

COMPANY: Delta Employee Credit Union
INDUSTRY: Financial services
FOCUS: Consumer banking

BUSINESS GOAL

Bolster customer service by enhancing account accessibility and funds management.

BACKGROUND/OPERATIONAL DESCRIPTION/RESULTS

The Delta Employee Credit Union has 30,000 customers, and more than half of them are on the go. The credit union's wireless service is an extension of its Internet banking services and allows customers to check account balances, transfer funds, and make payments anytime, anywhere. The credit union doesn't charge for the service, but instead sees it as a customer retention tool. It also plans to add a feature that will alert customers when their account balances are low. It took three months to develop the software, which includes a security encryption capability that prevents less secure devices from accessing the system.

TECHNOLOGY

The wireless service is available via digital cell phones or wireless PDAs. Delta worked with Air2Web Inc. to develop the middleware and the device applications that run on the selected hardware platforms. Users first connect with the Air2Web server, then via frame-relay to the Credit Union site, and then through a secure firewall for financial information.

Source: Beth Bacheldor, "Credit Union Flies Without Wires," *InformationWeek* (September 3, 2001): 56.

COMPANY: Dunkin' Donuts
INDUSTRY: Food services
FOCUS: Retail fast food

BUSINESS GOAL

Improve sales by pushing discount coupons to customers' mobile phones.

BACKGROUND/OPERATIONAL DESCRIPTION/RESULTS

Providence, Rhode Island–based Dunkin' Donuts operates thousands of corner donut shops in the United States plus in dozens of other countries around the world. Dunkin' Donuts Italy piloted a concept of advertising via wireless to boost sales. Customers would see a Dunkin' Donuts ad on a billboard or hear a message on radio, prompting them to call a number to download on their cell phone a digital coupon good for discounts on their next donut purchase. Users would call the number given, send a short message, and receive the electronic coupon as a reply. The technique leveraged the large base of mobile users in Italy while not spamming them. The two-month trial period resulted in a 9% increase in sales.

TECHNOLOGY

Dunkin' Donuts selected Mobileway to provide the software to manage the mobile data transactions, together with AdReact, a mobile marketing and advertising solutions provider.

Source: "Donut Sales Up, Thanks to M-Commerce Initiative," *M-Business* (February 13, 2002). Access at *www.mbusinessdaily.com.*

COMPANY: Equitable Insurance, Inc.

INDUSTRY: Insurance

FOCUS: Life insurance

BUSINESS GOAL

Sales-force automation and management

BACKGROUND/OPERATIONAL DESCRIPTION/RESULTS

Equitable is a large New York City–based insurance company with thousands of agents serving customers in every state. Equitable purchased 3,650 Pocket PCs for its salesforce to access the corporate network while on the road. Mobile salespeople can now access broker records, activity reports, and transaction histories, and synchronize their appointment calendars, contacts, and to-do lists with their corporate network.

The new system also provides a means to better measure and manage the sales force. "We get to know how many brokers they've [the salesforce] seen in a week and we can correlate that with sales," commented Equitable CIO Eric Jansen.

Bandwidth and coverage continue to be issues that limit the speed and convenience of the program, but there are benefits regardless, with even more envisioned when network enhancements are in place.

TECHNOLOGY

Compaq Ipaq Pocket PCs. San Diego–based Wireless Knowledge's Mobile Sales Desk software is integrated with Equitable's back-end systems. Middleware by Wireless Knowledge buffers data back and forth between the PDAs and the corporate databases.

Source: Paul Mcdougall, "Full-Fledged Business Apps Make PDAs Indispensable," *InformationWeek* (October 22, 2001).

COMPANY: Famous Footwear

INDUSTRY: Retail

FOCUS: Footwear

BUSINESS GOAL

Eliminate pricing errors in a rapidly changing price environment. Give sales associates a streamlined method of checking inventory and future specials to provide improved customer service.

BACKGROUND/OPERATIONAL DESCRIPTION/RESULTS

Madison, Wisconsin–based Famous Footwear is a $1.7 billion retailer that sells millions of pairs of shoes and employs some 10,000 sales clerks in its hundreds of stores countrywide. With prices changing frequently and advertised specials running almost every weekend, getting the price right in each store was important. Famous Footwear placed a WLAN access point in each store, linking it to the company's central system via a virtual private network. Then the company deployed handheld units with integral bar-code scanners, which connect to the corporate network to receive updated pricing information. Sales clerks scan a bar code for a particular shoe the customer has chosen, and the software displays the price, any discounts or sales, the shoe's availability both in store and in the network, and when the shoe may go on sale. Sales associates can now stay engaged with the customer, roaming from shoe to shoe, checking inventory and pricing while with the customer. The new system also has significantly reduced pricing errors. The company reported achieving a reduction in pricing errors approaching 75%.

TECHNOLOGY

Symbol SPT 1700 handhelds with integral barcode scanners

Source: *Business 2.0* (April 3, 2001); and *Wi-Fi@Work*, Synchrologic's online newsletter (April 1, 2002).

COMPANY: Fidelity
INDUSTRY: Financial services
FOCUS: Consumer brokerage

BUSINESS GOAL
Automate and improve customer access to account and trading information.

BACKGROUND/OPERATIONAL DESCRIPTION/RESULTS
Fidelity, the Boston-based provider of financial services, knew its customers were missing out on investment opportunities. Mobile customers were often away from hardwired PCs and didn't otherwise have access to market information. Fidelity began a multiyear program of developing mobile services for its customers. They began with a simple paging alert service called Instant Broker, which let active traders monitor their accounts through pagers. In 1999 Fidelity added two-way capability to allow transactions and expanded the information feeds. Today, Fidelity has expanded its wireless service even further into a product called Fidelity Anywhere, which allows users to manage their 401K accounts and much more. In 2001, Fidelity Anywhere had more than 92,000 registered users and was growing by 3,000 users per month—a small percentage of its total customer base, but an impressive percentage of its new-customer totals.

Fidelity created the position of Chief Wireless Officer and named Joseph Ferra to the position. Mr. Ferra reported that about one-third of Fidelity Anywhere's registered users represent new Fidelity accounts and that customer retention is high. Customers really like the convenience and control. Fidelity has realized benefits with each wireless initiative, including increased customer satisfaction (referrals and retention) and revenue (based on new accounts). Because Fidelity was one of the first to develop wireless capabilities in the financial market segment, they positioned themselves as part of the imbedded menus on many PDAs, a built-in primer for new customers!

TECHNOLOGY
Cingular Interactive network services; Certicom Corp. wireless products and services

Source: "The Wireless 25," *ComputerworldROI* (September 24, 2001).

COMPANY: First Service Network

INDUSTRY: Service

FOCUS: Building maintenance

BUSINESS GOAL

Reduce call center volume through wireless and Web-enabled self-service. Improve information flow to customers.

BACKGROUND/OPERATIONAL DESCRIPTION/RESULTS

Linthicum, Maryland–based First Service Network (FSN) is a nationwide maintenance service company that works with commercial building owners and retail stores to maintain their heating, air conditioning, and plumbing systems. FSN executes the work through 3,500 local specialty contractors located across the country. FSN takes the customer calls through a central customer service center, and then selects and dispatches a local contractor to do the work, following up with both the local contractor and the customer. The FSN call center was taking and making thousands of telephone calls per day.

In May 2001, FSN went live with its wireless notification system linked to the company's CRM applications. Under this system, dispatchers receive a customer's request for service, and then transmit the customer request, address, and customer history to the cell phone of a local contractor. Contractors must check in 15 minutes before scheduled arrival or the system automatically prompts the dispatcher to place a phone call to the contractor. Customers can monitor the status of their calls on the FSN Web site and can run reports on the nature and frequency of service calls to spot maintenance trends.

The new system is designed to provide customers with self-service on the Web site, freeing up as much as 80% of the call center staff. When customers fully adopt the self-service program, the 15-person call center staff will be able to more than triple its capacity.

TECHNOLOGY

Siebal Systems, J.D. Edwards back-office applications

Source: Robert Scheier, "The Wireless Workforce Pays Off," *M-Business* (October 2001): 37.

COMPANY: Florida Power & Light
INDUSTRY: Utility
FOCUS: Metering

BUSINESS GOAL
Improve management of field crews and streamline meter reading.

BACKGROUND/OPERATIONAL DESCRIPTION/RESULTS
Florida Power & Light (FP&L), headquartered in Juno Beach, Florida, has thousands of service crews criss-crossing the state to support the utility infrastructure. The daily challenge of dispatching and managing the service crews is compounded during a crisis, when efficient management of field crews becomes even more critical. FP&L invested $35 million upgrading its voice and wireless infrastructure and the communication systems in its mobile vans. Each van has global positioning equipment to help crews locate trouble spots and to help the dispatcher know exactly where the crew is located. The new system also tracks and manages what equipment crews have onboard. Ultimately, the new system will improve efficiency and customer service.

Another wireless innovation is the replacement of manual meter reading systems with wireless meter reading. Under an extensive pilot program, FP&L fitted thousands of FP&L customer meters with automated meter readers that transmit power consumption readings wirelessly. With the wireless system, more accurate readings can be taken more frequently. Meter readers don't have to enter a customer's property anymore. Data about customer power interruptions is transmitted instantly, 24 hours a day. The system has an immediate payback for larger commercial customers that read meters more frequently and for which accuracy is critical.

TECHNOLOGY
The wireless metering devices are from SmartSynch of Jackson, Mississippi, encased in devices made by Siemens. The meters transmit using Skytel's two-way paging system.

Source: Susannah Patton, "The Wisdom of Starting Small," *CIO* (March 15, 2001).

COMPANY: GE Medical Systems

INDUSTRY: Service and repair

FOCUS: Medical equipment

BUSINESS GOAL

Increase productivity of field service technicians. Improve timeliness of data reporting and updating of equipment status. Lower cost of hardware and support to field technicians.

BACKGROUND/OPERATIONAL DESCRIPTION/RESULTS

GE offers extended warranty and service on sophisticated medical scanning equipment to every major hospital in the United States. Providing prompt service and timely information updates to its customers is critical, in addition to economically and efficiently managing its service teams. GE calculated that its 2,400 field service technicians spend as much as one-third of their time on administrative tasks. Backlogs in paperwork often delayed prompt posting of equipment status and service call completion. Reporting to customers was not timely, reflecting poorly on GE. Therefore, GE sought to reduce administrative work time to 5% and to increase the timeliness of data from one week to four hours.

GE reengineered the entire reporting and communications process. GE service technicians carried laptops, but booting them up took too long. They switched to smart phones, which combine a digital phone with a Palm Pilot. GE also upgraded its service Web site and modified it to work with the Palm OS. GE technicians can now quickly log-on, enter and retrieve data, and link to FedEx's site to track parts deliveries.

The total cost of the project was $3 million, including Web applications to work with the smart phones and the hardware for the field technicians. GE calculates it will save $15 million annually by eliminating cell phones and laptops, replacing them with a single device.

TECHNOLOGY

Kyocera 6035 Smartphone with Palm operating system

Source: Howard Baldwin, "2001 Mobile Masters Awards for Enterprise Deployment," *M-Business* (January 2002).

COMPANY: Harris Bank

INDUSTRY: Financial services

FOCUS: Retail banking

BUSINESS GOAL

Expand customer service by providing customers with wireless banking services through mobile phones.

BACKGROUND/OPERATIONAL DESCRIPTION/RESULTS

Chicago-based Harris Bank, whose parent company is the Bank of Montreal, was the first U.S. banking institution to offer customers wireless banking services through handheld devices. The system allows customers to transfer funds between accounts, pay bills, view transactions in real time, and gain access to stock watch lists, news, and weather. The service is also available to all of the bank's customers; however, research shows that wireless services are apt to attract higher-value customers, which is a plus for any financial institution. In fact, the bank counts attracting high-value customers through a relatively low-cost wireless channel as one of the key business advantages of its wireless project.

Other advantages include the potential to expand service offerings beyond traditional banking and brokerage offerings. For example, the potential exists to provide value-added lifestyle and m-commerce services, to develop a mobile marketplace, and to resell wireless services.

TECHNOLOGY

Toronto-based 724 Solutions' Internet wireless banking infrastructure software is accessible by a wide range of Internet access devices, including Personal Communications Services (PCS) phones or 3Com Palm computing devices.

Source: Alan Radding, "Leading the Way on Wireless," *ComputerworldROI* (September 2001).

COMPANY: Illinois State Police
INDUSTRY: Government
FOCUS: Law enforcement

BUSINESS GOAL

Improving the coverage of existing telecommunications capability while adding remote and rapid access to databases

BACKGROUND/OPERATIONAL DESCRIPTION/RESULTS

With more than 3,000 officers and civilians, the Illinois State Police (ISP) force is responsible for providing law enforcement throughout the entire state, often supporting small and remote communities with backup policing. Their legacy radio communications systems were more than 20 years old, and coverage was limited to major metropolitan areas. Their goals in deploying a new wireless mobile computing and communications system were straightforward: seamless statewide coverage, immediate access to information in key law enforcement databases, and ease of use for officers in the field.

The ISP chose the CDPD option for data transmission, which leverages the unused bandwidth on the analog channels of wireless phone networks maintained by AT&T and Verizon. CDPD runs on existing cellular towers offering only 19.2kbps speed but low cost and reliability. The ISP saved millions by having to maintain its own network. Security was handled by several levels of encryption. Patrol cars were outfitted with rugged laptops and higher-powered wireless modems. Most of the software was off the shelf, including the barcode reader for quickly scanning in drivers' licenses and a small thermal jet printer to receive low-resolution black-and-white photos of suspects. The units also have GPS capability, allowing tracking and dispatch of cars by central and emergency locater help.

TECHNOLOGY

Laptops are Panasonic CF27, thermal printer by Pentax PocketJet II, barcode reader by Symbol 2D and image capture device. Wireless modems are Spider 4 units by Enfora.

Source: Steve Barth, "Calling All Cars," *Field Force Automation* (October 2001).

COMPANY: Interbrew (Belgium)
INDUSTRY: Food service
FOCUS: Beverage

BUSINESS GOAL

Improve productivity of 160 technicians by reducing paperwork, phone calls, and office visits, allowing them to make more service calls. Improve recordkeeping. Interface with existing SAP system.

BACKGROUND/OPERATIONAL DESCRIPTION/RESULTS

The world's second largest brewer, Interbrew sells beer in more than 110 countries. To insure high quality, a team of 160 technicians visits pubs and hotels once every eight weeks to clean beer-dispensing system tubing and to provide maintenance and repairs on demand. Before the new wireless system, all recordkeeping was paper-based. Ordering parts was performed via phone, and following up on orders was time consuming.

The new system allows field technicians to see customer requests for service immediately on their handheld units, as well as update their schedules, place orders, receive messages from the customer service group, e-mail other technicians for advice or help, and transmit completed work orders for prompt customer billing.

Service technicians now make more service calls per day and management can track productivity better.

TECHNOLOGY

DatAction and Idesta BV, together with wireless network provider RAM Mobile Data, created a custom solution: Intermec 6110 handheld computers running on windows CE. Vehicles were equipped with RF modems for wireless transmission.

Source: "Handhelds Help Keep the Beer Flowing," case study, *Pen Computing* (October 2001).

COMPANY: Intercoastal Realty
INDUSTRY: Real estate
FOCUS: High-end home sales

BUSINESS GOAL

Provide agents and clients with anytime, anywhere access to listed home inventory information and images.

BACKGROUND/OPERATIONAL DESCRIPTION/RESULTS

Intercoastal Realty is a Florida-based real estate company specializing in high-end homes. Previously, Intercoastal had incorporated virtual tour capabilities of homes listed for sale on its Web site, as had several other area real estate firms; however, the images and information could only be viewed from hardwired computers, so if agents traveling with a client spotted an interesting listing, they would have to call for a viewing or return to the office to gather information and images. With their new wireless system, agents can access color images of homes and the 360-degree virtual media tour on their Pocket PC–based handhelds, and show them to clients on the spot.

Intercoastal predicts the new system will allow clients more choices more quickly, speeding the time it takes to review available homes and ultimately closing real estate deals more quickly.

TECHNOLOGY

Handhelds use the Microsoft Pocket PC software with UR There's 360-degree Virtual Media Tour technology modified for use in handhelds.

Source: Microsoft case studies: Intercoastal Realty (*www.microsoft.com/resources/casestudies/company.asp*); and *www.intercoastalrealty.com*.

COMPANY: Jim Hudson Lexus/Saab
INDUSTRY: Automotive
FOCUS: Retail

BUSINESS GOAL

Provide timely business alerts to the salesforce as a way of improving response times to customers.

BACKGROUND/OPERATIONAL DESCRIPTION/RESULTS

Consumers are increasingly using the Web to shop for cars. When they are ready to get a quote from a dealer, they select the dealers and send an electronic request for a quote. Jim Hudson Lexus/Saab, a Columbia, South Carolina–based dealer, was receiving increasing numbers of Internet requests for quotes. They quickly learned that the first dealer to respond to the Internet consumer had a better chance of building the personal relationship required to close the sale.

To improve its response time, Hudson equipped one sales manager at each location with a Research in Motion BlackBerry wireless e-mail pager to monitor information requests from Internet customers. Internet sales manager Tony Albanese, who also carries a BlackBerry, indicated it is a great help to ensure that he can respond to a customer's inquiry while that customer is still on the Web. Lexus now mandates that dealers must respond to customer e-mails within minutes, not hours, explained Albanese. For Hudson, sales through the Internet are up 15%.

TECHNOLOGY

BlackBerry wireless e-mail pager with RIM Web server application

Source: Paul Mcdougall, "Full-Fledged Business Apps Make PDAs Indispensable," *InformationWeek* (October 22, 2001).

COMPANY: JS Express
INDUSTRY: Transportation
FOCUS: Courier

BUSINESS GOAL

Enhance customer service by automating and streamlining the dispatch process and better managing drivers.

BACKGROUND/OPERATIONAL DESCRIPTION/RESULTS

JS Express is a $25 million, St. Louis–based courier company whose 125 drivers deliver more than 3,000 packages daily. Their promise is to pick up within 15 minutes of a customer's call and to complete delivery within an hour of the call. On the outside, all was well, but inside the JS dispatch office, it was chaos. Ten dispatchers reading from paper slips barked delivery details and addresses in rapid fire to drivers over a two-way radio system. The drivers often got the details wrong, so they began carrying tape recorders! Channel clutter and user overload often resulted in confusion. Dispatchers would then pass written notes to customer service coordinators, who would call customers to confirm their orders and pickup times based on their best estimate of when the driver got the verbal order.

JS spent nearly two years working with a vendor to customize software to meet the company's requirements for a high-volume, real-time environment. The new system assigns the drivers based on current location and customer requirements. It also arranges the order of pickups and deliveries to correspond to their location. Both phone and Web orders are entered into the system. The system calculates the estimated time of arrival and posts it for customer service personnel to use in communicating with customers. Drivers post to the system via WAP phones as they complete a pickup, and the system updates the records instantly.

The new implementation was not without bugs. Surprisingly, JS made a major change after the deployment, switching from the pricey handheld units initially deployed to simple WAP phones on more reliable networks. Drivers access the Web server for their next assignment, instead of being pushed the information. JS opted for low cost and simplicity to lower the total cost per driver.

JS spent more than $2 million for the new system and believes that productivity has significantly increased and costs have decreased. Since the new system was fully implemented, volume has increased from 3,000 deliveries per day to 6,000, as the number of dispatchers went from 10 to 5.

TECHNOLOGY

The customized software program was developed by Cheetah Software Systems, Inc. in Westlake Village, California.

Source: Merrill Douglas, "The Route to Success," *Field Force Automation* (July 2001).

COMPANY: McKesson Corp.

INDUSTRY: Wholesale distribution

FOCUS: Pharmaceuticals

BUSINESS GOAL

Improve warehouse efficiency and reduce errors.

BACKGROUND/OPERATIONAL DESCRIPTION/RESULTS

McKesson is the nation's largest wholesaler of drugs. From aspirin to Pepto Bismol, the company's warehouse stocks tens of thousands of products. In a single shift, a picker can track down and pull as many as 4,000 products. The paper required to track the process was extensive and laced with human errors. This resulted in mis-placed products, inaccurate inventory, and delays in orders. Based on industry data that estimate the total cost of an error to be around $100 (which covers the cost of initial shipping and processing the return), McKesson made the leap to automate the process. Now pickers wear wireless computers strapped to their wrists—quite a leap for the 168-year-old company.

Receiving data over a wireless network inside each warehouse, the wrist computers display the detailed order information and the exact row, shelf, and bin where the product is located. The system will route the picker in the most efficient sequence through multiple-order picking. Pickers pull the items from the shelves and scan the product barcodes with a scanner on their fingers worn like a ring. Occasionally, the system will run cycle counts while pickers are picking, asking them "How many boxes are left in the bin?" This information is then entered on the wrist device and incorporated into the warehouse inventory system.

McKesson spent $52 million to outfit 1,300 warehouse workers with the wearable wireless devices ($3,000 each) as well as equipping its 52 warehouses with wireless local area networks. Tom Magill, former VP of Logistics Technologies for McKesson, stated that the system has already saved millions and would pay for itself many times over. McKesson showed an 8% gain in productivity, an 80% drop in incorrect items shipped, and a 50% drop in product shortages. Inventory accuracy has improved to 99.5%.

McKesson also extended the system to their delivery and logistics operations (see additional case study).

TECHNOLOGY

AvantGo, Inc., M-Business server software integrated on Symbol SPT 1700 hand-held devices.

Source: Matthew G. Nelson, "Easy-To-Track Deliveries," *Information Week* (September 10, 2001); and "Doing Business Without Wires," *Information Week* (January 15, 2001): 22–24.

COMPANY: McLane Co.

INDUSTRY: Distribution

FOCUS: Grocery

BUSINESS GOAL

The more efficient management of a large fleet of trucks carrying perishable goods. Automate the entry and retrieval of shipping information to provide more access into the distribution process.

BACKGROUND/OPERATIONAL DESCRIPTION/RESULTS

McClane is a Temple, Texas–based grocery delivery business with a fleet of 1,050 trucks and 17 distribution terminals. Before the new system, the primary communication methods used to stay in touch with more than 1,000 drivers were pagers, dispatchers, and a lot of paper. With the new system, every driver carries a handheld unit equipped with a barcode scanner. Drivers use the device to record delivery information and capture electronic signatures. When a shipment is complete, drivers dock their units in a cradle located in the cab of the truck, which is part of an onboard communications systems that uses a dual-mode system comprising a wireless LAN and satellite link to upload the information to McLane's central system. The system cost approximately $15 million, with an estimated payback of two years, according to Dave Dillon, Manager of Transportation at McLane.

TECHNOLOGY

Symbol Technologies' handheld devices with track-and-trace software, as well as signature capture software by IBM. A satellite link is provided by Qualcomm.

Source: Bob Brewin, "Trucker McLane Rolls Out Dual-Mode Wireless Vehicle System," *Computerworld* (May 1, 2001).

COMPANY: Mesa Energy Systems
INDUSTRY: Service
FOCUS: HVAC

BUSINESS GOAL

Improve communications to field service personnel. Automate and streamline time reporting and billing.

BACKGROUND/OPERATIONAL DESCRIPTION/RESULTS

Mesa Energy Systems, based in Irvine, California, is a full-service heating, ventilating, and air-conditioning service company operating in San Francisco, Los Angeles, and San Diego. Mesa wanted a better way than radio communications and a paper-based billing process to communicate with and manage its extensive field service force. Mesa chose a mobile computing solution with wireless capability.

The new system included software applications for time sheets, electronic work orders, equipment inventory, signature capture, and a mapping program to help find customer locations faster. Field technicians can access the building history file and service history on individual pieces of equipment within the building. When a work order is completed, the customer signs off on the unit and the order information is transmitted back to the central office to generate a bill. Customers are faxed their completed work orders within minutes of job completion and order closure. The application also includes e-mail for technicians to use to stay in touch with other technicians and the central office.

Better access to details on open jobs has improved the billing cycle and cash flow. Faster response times to customers on the initial dispatch, as well as detailed reports after the service call, have both increased customer satisfaction and retention. Mesa estimates a 15% increase in its field force productivity, as well as a reduction of dispatcher time. As an additional benefit, Mesa tested sending a wireless communication to each field technician, prompting them to ask customers for more business in various other service areas. Using the prompts, service technicians flooded the company with new leads at a rate of 10 to 15 per day versus the previous 8 to 10 per week, without the reminders.

TECHNOLOGY

Handhelds are rugged Itronix T5200 handheld PCs. Customized service operations software is by FieldCentrix.

Source: Microsoft case studies: Mesa Energy Systems (*www.microsoft.com/resources/casestudies/company.asp*). See also, "The Wireless 25," *ComputerworldROI* (September 2001).

COMPANY: Miami Dade Water and Sewer

INDUSTRY: Municipal government

FOCUS: Water and sewer

BUSINESS GOAL

Improve the accuracy of system maps, allowing field crews to more rapidly locate service location points.

BACKGROUND/OPERATIONAL DESCRIPTION/RESULTS

The Miami Dade Water and Sewer (MDWASD) system is the largest in the southeast and includes more than 414 square miles of water and wastewater pipes, connections, manholes, fire hydrants, and other equipment. Accurately mapping and updating the various flow points, level indicators, and other structures to support efficient dispatching of hundreds of field crews was challenging. Outdated paper-based maps did not accurately reflect the rapid growth in infrastructure, nor did it allow prompt updating. Field crews would waste valuable time sorting through hundreds of paper maps while in the field.

MDWASD used GPS technology coupled with digital mapping to digitize the entire infrastructure. The first task was to accurately map the thousands of control devices, valves, fire hydrants, and access points. Field technicians would use the GPS units to record location information to a high degree of accuracy and then upload the information to a computer-aided design (CAD) mapping system. In 18 months the teams converted more than 1,100 paper maps, identifying 170,000 locations to digital format. Field crews now locate structures and control points using wireless pen-based computers linked to the central digital mapping files.

TECHNOLOGY

GPS and mobile data hardware support provided by Trimble—the pathfinder GPS receiver and ASPEN field-data collection software. Pen tablets were furnished by WalkAbout Computer, the hammerhead unit.

Source: "The Wireless 25," *ComputerworldROI* (September 2001); and Gary Thayer, "Handhelds Help Miami-Dade Water Tackle Multi-Million Dollar GIS Project," *MobileVillage.com* (February 15, 2002).

COMPANY: Moses Cone Healthcare System
INDUSTRY: Health care
FOCUS: Hospital patient records

BUSINESS GOAL

Improve efficiency and reduce errors by optimizing patient record retrieval and updating and the transmission and fulfillment of prescription medicines.

BACKGROUND/OPERATIONAL DESCRIPTION/RESULTS

Moses Cone Healthcare System, a large regional health care center located in Greensboro, North Carolina, has deployed a wireless electronic patient records system. Doctors, pharmacists, and physician assistants can access hospital records and up-to-the-minute lab results using handheld PDAs at any of their four hospital locations. The handhelds are equipped with infrared ports, enabling clinicians to access patient records and download treatment plans from synchronization sites located around the hospital. Doctors can regularly update their PDAs by synchronizing them with the hospital system throughout their shifts.

John Jenkins, the VP and CIO of Moses Cone, commented that the system cost around $250,000 to install, an investment that he expects will easily be recovered in error reduction and a significant time savings for doctors. The system eliminates the need for doctors to sit down at workstations to review patient records or complete and print out required forms. Doctors say that the system saves them anywhere from 30 minutes to an hour per day. Pharmacists also save time because prescriptions arrive via the system, instead of on paper. Data about drug interactions can be quickly pulled down and referred to in patient instructions. In addition to staff physicians, doctors who practice in the region are allowed access to the system.

TECHNOLOGY

MercuryMD Inc.'s Staging Database and Extended Systems Inc. XTNDConnect Server. Palm IIICs or any Palm OS device with infrared port. Clarinet synchronization towers on each floor.

Source: Mathew G. Nelson, "Doctors Trade Clipboards for PDAs," *InformationWeek* (July 30, 2001); and "Medical Professionals Manage Patient Data with Handhelds," *www.mobilevillage.com* (February, 6, 2002).

COMPANY: Owens & Minor, Inc.

INDUSTRY: Distribution

FOCUS: Health care

BUSINESS GOAL

Optimize the process of replenishing and billing materials placed in remote customer storage points as well as tracking consigned inventory.

BACKGROUND/OPERATIONAL DESCRIPTION/RESULTS

Richmond, Virginia–based Owens & Minor (O&M), a $3.8 billion distributor of medical supplies to hospitals and clinics, works with its customers to reduce their onhand inventory by restocking supplies on a daily or weekly basis as needed. O&M has equipped the operating rooms of its customers with a predetermined level of surgical supplies; each item is labeled with a barcode. When surgeons use an item, a wireless handheld is used to scan the item, and then the scanner transmits the consumption information to O&M, which adjusts the inventory, bills the customer, and sets up the reorder.

TECHNOLOGY

Information not disclosed

Source: *InformationWeek* (September 17, 2001): 140.

COMPANY: Packaged Ice, Inc.

INDUSTRY: Manufacturing

FOCUS: Retail sales

BUSINESS GOAL

Improve equipment performance and customer service; cut costs by reducing need for onsite service.

BACKGROUND/OPERATIONAL DESCRIPTION/RESULTS

Packaged Ice, Inc., a Houston–based provider of ice machines and ice to more than 2,000 locations across the United States now uses wireless to monitor and correct problems in the machines. Before using the wireless technology, a service technician would be dispatched every time a machine would have a problem. Packaged Ice installed a wireless modem in every machine and connected it to Web-based remote maintenance software.

The new system monitors more than 100 physical conditions, such as the ice production, temperature, and machine door status. When the machines transmit information about a problem, an operator can diagnose and attempt remote fixes, such as resetting the computer or clearing the loading tray. About one-quarter of the machines that break can be fixed remotely, without dispatching a service technician. Maintenance costs have dropped since Packaged Ice began remotely sensing its installed base of ice machines; the annual cost of the wireless modems is less than $120 per machine. The savings are significant and estimated by Packaged Ice at $560,000 per year.

TECHNOLOGY

Remote maintenance software by Isochron Data, Austin, Texas

Source: "The Wireless 25," *ComputerworldROI* (September 2001).

COMPANY: Pfizer, Inc.

INDUSTRY: Pharmaceuticals

FOCUS: Manufacturing and shipping

BUSINESS GOAL

Cut costs, improve manufacturing operations.

BACKGROUND/OPERATIONAL DESCRIPTION/RESULTS

Pfizer's eight-story plant in Brooklyn, New York, contains a plethora of wireless technologies, all aimed at reducing and/or eliminating paper, streamlining processes, and increasing efficiency while cutting costs. Thirty-five radio frequency (RF) antennas support the wireless transfer of process, inventory, and other information from Palm devices and barcode scanners to central systems. For example, a wireless terminal is used to present the bill of material to an operator. The operator uses this information to dispense and subdivide components. The system provides real-time work-in-progress tracking, with all material moves recorded into a central database. Plant floor processes such as this one once required tons of paper. Now wireless applications monitor manufacturing processes and progress in real time, eliminating clerks carrying clipboards and checking off items on paper for subsequent reentry into a computer system. Thomas J. Cala, team leader of enterprise systems at Pfizer, says wireless technologies have, among other things, greatly reduced data entry requirements, which alone has saved the company millions of dollars.

TECHNOLOGY

Real-time data entry via barcode scanning using Symbol SPT 1740 running Palm operating system; 802.11 wireless LAN technology.

Source: "The Wireless 25," *ComputerworldROI* (September 2001).

COMPANY: PRI Automation
INDUSTRY: Manufacturing
FOCUS: Semiconductor equipment

BUSINESS GOAL

Improve manufacturing process by enabling workers to access a manufacturing knowledge management system on an as-needed basis, anytime, from anywhere.

BACKGROUND/OPERATIONAL DESCRIPTION/RESULTS

Bellerica, Massachusetts–based PRI Automation commissioned software vendor Generation 21 to build a Web-based training and support system that supplies PRI workers with information on everything from routine system maintenance to mission-critical equipment error recovery. It then equipped 30 workers with wireless access to the system as part of a test. The test revealed that workers could use richer access to multimedia, so PRI migrated the system from Palm handheld devices to full-featured PC tablets equipped with wireless networking cards.

Employees are now able to access information they need, when and where they need it, without needing to go through cumbersome paper-based reference books. PRI plans to extend the system to its clients, allowing customers to perform maintenance and repair checks, parts management, and other tasks from their own sites.

TECHNOLOGY

The tablet PC can be used either with a wireless network card (Cisco Airnet) connected to a local transmitter or with a stand-alone interface card (BlackBerry), allowing external reception.

Source: Alan Radding, "Leading the Way with Wireless," *ComputerworldROI* (September 2001).

COMPANY: Producers Lloyds Insurance Co.

INDUSTRY: Insurance

FOCUS: Crop insurance

BUSINESS GOAL

Improve customer service at the point of sale.

BACKGROUND/OPERATIONAL DESCRIPTION/RESULTS

As a leading Texas-based crop insurer, Producers Insurance agents are in the field most of the time, working with customers to assess, price, and provide insurance coverage for crops in various stages of development. Farmers need timely quotes to cover crops, and Producers' agents must access corporate information to develop a quote. Previously, agents would spend time on the phone working through the details or have to return to the local office to prepare a quote for later delivery to the farmer.

Producers deployed a wireless system to empower its field agents to service customers immediately. With the new system, Producers agents access the corporate database wirelessly for pricing information that corresponds to crops, timing, and the amount of coverage. With all of the information at the point of need, Producers' agents can make more accurate decisions while in the field, supporting customers on the spot with accurate quotes and confirmed coverage. Benson Latham, Producer's VP of marketing, commented that providing all of the needed information to the agent, including weather forecasts, improves decision-making and customer service.

Producers has invested around $100,000 in the project and believes a quick return is possible based on additional business closed.

TECHNOLOGY

Wireless technology from Emrys Technologies, Ltd.

Source: "The Wireless 25," *ComputerworldROI* (September 2001); and Alan Radding, "Leading the Way on Wireless," *ComputerworldROI* (September 2001).

COMPANY: St. Luke's Hospital
INDUSTRY: Health care
FOCUS: Patient care

BUSINESS GOAL
Improve patient care by automating patient monitoring and check-in.

BACKGROUND/OPERATIONAL DESCRIPTION/RESULTS
When patients check-in to St. Luke's Hospital in Houston, Texas, they get a typical hospital wristband but with one addition—a barcode. The barcode links the hospital staff to the electronic patient records system, which contains each patient's entire file. Whenever the patient is treated or medicated, the barcode is scanned and the records are updated.

Hospital staff use small wireless laptops with wireless network interface cards. Multiple staff members and doctors can access the patient records simultaneously. St. Luke's spent around $1 million on the system and estimates a 15% to 20% increase in staff productivity due to the new system. Significant error reduction is also an anticipated benefit of the new system.

TECHNOLOGY
CISCO AiroNet 802.11b network

Source: *Wi-Fi@Work*, Synchrologic's online newsletter (April 1, 2002).

COMPANY: Scandinavian Airlines System
INDUSTRY: Airlines
FOCUS: Passenger travel

BUSINESS GOAL

Improve the dissemination of rapidly changing flight schedule and crewing information to crews on a 24/7 basis. Provide productivity tools to mobile employees.

BACKGROUND/OPERATIONAL DESCRIPTION/RESULTS

Scandinavian Airlines System (SAS) is a consortium of three airlines flying to 100 destinations in 31 countries and serving more than 22 million passengers per year. SAS needed to automate and mobilize the relaying of flight schedule and crewing changes to their flight crews, who are constantly on the move, as well as provide their employees with productivity tools such as remote e-mail and calendar access.

The new system gives employees wireless access through GSM modules on their handhelds and through infrared docking stations at terminal locations. With more than 7,000 handhelds deployed, SAS is in much better communication with its workforce and can redeploy personnel with the confidence that new scheduling information will reach the employees and be confirmed back via e-mail.

TECHNOLOGY

Compaq Ipaqs running Windows Pocket PC

Source: Microsoft case studies: Scandanavian Airlines System (*www.microsoft.com/ resources/casestudies/company.asp*).

COMPANY: Starbucks
INDUSTRY: Retail
FOCUS: Beverage

BUSINESS GOAL

Improve store traffic and revenues by offering customers the convenience of wireless high-speed Internet connections.

BACKGROUND/OPERATIONAL DESCRIPTION/RESULTS

Starbucks, the leading coffee brand and coffee retailer in the United States, decided that installing WLANs in more than 70% of its 4,000 stores by 2003 made perfect sense. Research shows that 90% of Starbucks customers are Internet users, and often they are equipped with Internet-capable PDAs and laptops. Starbucks is betting that customers who come in for coffee will stay longer, come more often, and buy more while checking their e-mail or taking care of business.

With the new WLANs, customers can be linked to the Internet without unattractive wires and jacks cluttering the store. Darren Huston, Starbucks senior VP for new ventures, clarified that Starbucks wasn't interested in getting in the Internet access business, but in selling more coffee. Starbucks believes that the high-speed wireless Internet connections are one more reason to stop in at Starbucks.

TECHNOLOGY

A cooperative partnership was formed with Compaq, Microsoft, and MobileStar Network to quickly roll out the WLANs.

Source: John Markoff, "Starbucks and Microsoft Plan Coffeehouse Web Access," *New York Times* (January 4, 2001); And George Anders, "Starbucks Brews a New Strategy," *FastCompany* (August 2001).

COMPANY: Trinity Development
INDUSTRY: Construction
FOCUS: Highway

BUSINESS GOAL
Shorten cycle time in materials receiving and billing. Improve cash flow.

BACKGROUND/OPERATIONAL DESCRIPTION/RESULTS
Trinity builds highways in Ohio. After materials were delivered to a remote job site, it would take days or sometimes weeks for receiving paperwork to make it back to Trinity's office so that Trinity could bill the customer. Trinity sought to speed the overall process while eliminating paper.

Using wireless handheld devices, field personnel now key in the receiving information for materials delivered and transmit it directly to the Trinity billing system, which produces a bill to the state of Ohio that same day. Trinity is receiving payments from the state weeks faster than prior to the wireless system. This has allowed the company to save the 3% of its revenue that used to go toward interest payments.

Often, Trinity can pay its, suppliers in advance, earning a 2.5% discount. The savings on millions of dollars' worth of materials per year has been significant, enabling a six-month payback on the wireless investment. Trinity's President Jim Wilson stated, "It has been miraculous."

TECHNOLOGY
Windows CE–based handheld computers and software from Clayton IDS.

Source: *M-Commerce* (March 6, 2001).

COMPANY: Vail Resorts
INDUSTRY: Recreation
FOCUS: Skiing

BUSINESS GOAL

Improve the accuracy of scheduling and employee utilization.

BACKGROUND/OPERATIONAL DESCRIPTION/RESULTS

Vail Resorts is a premier ski destination consisting of four resorts covering 4,000 acres of skiable terrain and the largest ski school in the world. A staff of 1,000 ski instructors delivers more than 25,000 skiing lessons each season. Before the wireless system, instructors carried paper notes and schedules in their pockets listing guests, schedules, meeting locations, and details of the lessons. Guests would request lessons in advance and all during the day. Ski school coordinators would attempt to match guest needs, such as snowboarding, Nordic style, or beginning instruction, with instructor skills; however, once instructors were on the slopes for the morning lessons, making changes during the day was an uphill battle.

With the new wireless system, instructors carry wireless handhelds with schedules and guest lists. A WLAN is in place, with nodes placed throughout the resort, allowing instructors to download new information and schedule changes. Vail developed the software inhouse in about a month using two programmers. Supervisors can locate the most appropriate and available instructor from a master list, then drag and drop the name into the scheduling routine. Instructors like the system because they are always aware of their schedules, making it easier to be prompt with guests.

TECHNOLOGY

Handhelds are HP 360 with Proxim wireless LAN adapter. The handhelds run Windows CE connected to a Microsoft SQL Server database.

Source: Microsoft case studies: Vail Resorts, Inc. (*www.microsoft.com/resources/casestudies/company.asp*); and Matt Hanblen, "This LAN Works on the Slopes," *Computerworld* (November 16, 1998).

COMPANY: Visiting Nurses Association
INDUSTRY: Health care
FOCUS: Home nursing

BUSINESS GOAL
Improve productivity of mobile nurses by streamlining patient recordkeeping.

BACKGROUND/OPERATIONAL DESCRIPTION/RESULTS
A shortage of nurses prompted the Visiting Nurses Association (VNA) of Santa Ana, California, to look for a way to reduce paperwork and allow nurses more time to see patients. According to a VNA analysis, every hour of health care delivered means 30 to 45 minutes of paperwork to document and bill. The average mobile nurse visits between five and eight patients per day, and busy nurses would need to take files home in order to stay current.

The VNA decided to write its own application for the Palm OS. The first applications it developed were to document each visit and time and attendance reporting. As the VNA became more adept at using the forms-based application development tools, they developed applications for scheduling, supplies ordering, infection control, and drip calculators. The applications are forms-based and provide lists to select from that allow easy completion. The VNA also incorporated a drug interaction database that holds data on more than 460,000 drugs. Patients average 10 medications each and often are receiving drugs from several physicians who may not be aware of each other.

Each day nurses upload that day's patient list, schedule, time and attendance records, patient demographics, and physician orders. When traveling, the nurses' handhelds can be connected via wireless or synchronized at night when they return home. The synchronization updates the medical records with that day's visit information. More than 100 nurses are using the system. Paperwork has decreased by 50% or about three hours per nurse per day.

TECHNOLOGY
Palm OS, Pendragon Forms, from Pendragon Software Corp.

Source: *Pen Computing* (November 2001): 32.

COMPANY: Volkswagen

INDUSTRY: Manufacturing

FOCUS: Auto

BUSINESS GOAL

Improve security, increase efficiency of inventory management and shipment preparation operations.

BACKGROUND/OPERATIONAL DESCRIPTION/RESULTS

Volkswagen places a Radio Frequency Identification Device (RFID) on the windshield of every finished and nearly finished automobile rolling off its assembly line at plants in Germany. The cars are parked in massive lots, which are patrolled by security guards in golf carts. These officers are equipped with RFID readers and notebook computers linked to Volkswagen's inventory management system. The RFID tags store information on the current condition of the car, remaining accessories to be installed, and additional prep work required before shipment. This information is relayed in real time to the inventory management system for accurate management of vehicles as they are prepared for loading and shipment to dealers worldwide.

TECHNOLOGY

Intelligent Long Range RFID from Identec Solutions

Source: Steve Konicki, "Sophisticated Supply: The Fast Track: Radio-frequency devices promise to make it easier to monitor the flow of inventory across the supply chain," *Informationweek* (December 10, 2001).

Appendix C

Resources on the Web

www.agilent.com/cm/wireless/dictionary/a.html
Agilent Technologies Wireless Dictionary

www.cnp-wireless.com/links.html
Even more links to wireless-related sites

www.computerworld.com
Wireless knowledge center

www.easyreservations.com
A wireless service for reserving flights, hotels, and automobiles

www.microsoft.com/isn/ind_solutions/wireless_data.asp
Microsoft's wireless data page

www.mobilecomputing.com
Mobile Computing's online magazine

www.umts-forum.org/glossary.html
The UMTS Forum's Glossary

www.vindigo.com
A web service for PCs or Palm PDAs. Tell it where you are or where you are going, and it finds the nearest and best places to eat, shop, and play.

www.wayport.net
Wayport offers wireless Internet access in airports and hotel public areas, and wired access in hotel guest rooms and meeting rooms.

www.wired.com/news/wireless
Wired magazine's wireless site

www.wireless.com
A site for everything wireless

www.wirelessweek.com
The Wireless Week Web site provides breaking news and examines the issues facing today's wireless industry.

Glossary

802.11

A family of wireless specifications developed by a working group of the Institute of Electrical and Electronics Engineers (IEEE). These specifications are used to manage packet traffic over a network and to ensure that packets do not collide, which could result in loss of data, while traveling from their point of origin to their destination (i.e., from device to device).

3G

Third generation. An industry term used to describe the next, still-to-come generation of wireless applications. It represents a move from circuit-switched communications (where a device user has to dial into a network) to broadband, high-speed, packet-based wireless networks (which are always on). The first generation of wireless communications relied on analog technology (see analog), followed by digital wireless communications. The third generation expands the digital premise by bringing high-speed connections and increasing reliability.

Adaptive Frequency Hopping

A method whereby a Bluetooth radio would first check that a band was clear before it attempted a transmission. This would allow Bluetooth radios to peacefully exist better with other radios such as 802.11 lb.

Advanced Mobile Phone Service

See AMPS.

AMPS

Advanced mobile phone service, commonly known as analog cellular. AMPS service is available in the United States, Mexico, Canada, Australia, and several other countries. It is the most basic wireless phone service. Uses CDPD format to transmit data.

Analog

A signal that streams continuously and varies in strength versus a digital signal that is a series of on and off signals.

ASR Automatic speech recognition. The technology that recognizes and interprets voice commands from a user to form electronic messages.

Automatice Speech Recognition See ASR.

Bandwidth A measure of the capacity of a communications channel and the amount of frequency available to a system. The wider the bandwidth allocated to a channel, the greater the data rate for a given protocol.

Biometrics Using biologic characteristics of an individual to authenticate identity via a device that scans and interprets the biologic data and encodes it electronically. Examples include retinal scans, fingerprint scans, and dynamic signature analysis. An intelligent pen device with imbedded sensors captures the characteristics of a person's signature, such as writing speed and pressure, and then uses the profile to authenticate the user's signature on subsequent occasions.

BlackBerry Device A corporate e-mail pager. The basic model is similar in size to a pager, designed for e-mail receiving and sending. This device is continuously connected, allowing the user immediate access. The device uses a new style of microkeyboard named Qwerty, which has prompted users to learn a double-thumb typing style on the tiny keyboard. The device is manufactured by Research in Motion (RIM) and although narrow in application (e-mail only) it is extremely usable and reliable. E-mails are forwarded through the RIM servers but are compatible with Microsoft Exchange servers and others. As the popularity grows, so does the loading on the pager network. Some major markets experience congestion, so check it out in advance. Also check out the monthly fees.

Bluetooth A short-range wireless specification that allows radio connections among devices within a 10-meter range of each other. Bluetooth is designed as a personal area network (PAN) technology with a wide variety of theoretical uses, although few products have been released that incorporate the technology, such as printers, phones, and laptops without wires. The vision was for two people to enter a room and immediately be able to transfer data between

devices on a common channel, or for you to walk into a
room with your laptop and be able to print on the nearest
printer without plugging into the network.

CDMA   Code division multiple access. A 2G standard used in the
United States by Sprint and Verizon. Also used in
Canada, Australia, and some southeastern Asian countries
(e.g., Hong Kong and South Korea). CDMA transmits
voice and data over the air by assigning users digital
codes within the same broad spectrum. Advantages of
CDMA include higher user capacity and immunity from
interference by other signals.

CDMA2000   The 3Glite (higher speed) communication standard based
on CDMA protocol; also known as 1XRTT. Offered by
Sprint and Verizon, it is backward compatible with
CDMA and thus will be easier to roll out city by city. It
will effectively double a CDMA carrier's voice capacity;
however, it transmits data a bit slower than WCDMA.

CDPD   Cellular digital packet data, a protocol used over an analog
wide-area network (AMPS analog cellular networks).
CDPD piggybacks data onto cellular analog conversations
to enable simultaneous voice/data transmission.

Cell   The basic geographic unit of a cellular system and the
basis for the generic industry term *cellular*. A city is divided
into small cells, each of which is equipped with a low-
powered radio transmitter/receiver or base station. The
cells can vary in size depending on terrain and capacity
demands.

Cellular Digital   See CDPD.
Packet Data

Circuit Switched   A classification for networks where the device connects to
the network only when placing or receiving a call, such as
with a traditional phone line. Next-generation wireless
networks will use packet-based networks, which are
always connected.

Code Division Multiple Access   See CDMA.

CRM   Customer relationship management. CRM programs are software and processes that help create an integrated information model used for planning, scheduling, and controlling all of the presale, selling, and postsale activities in the organization. Companies use a variety of methods to achieve a cohesive CRM approach, including buying expensive software and even more expensive consultants to help them do with more precision what they were doing for the most part anyway. Most companies have many of the informational components for CRM ready to extend; they simply must organize and focus the information.

Customer Relationship Management   See CRM.

Dual Band   Dual-band phones are capable of using two different frequencies of the same technologies. For example, a TDMA or CDMA phone that can use either the 800 or 1,900 MHz band. There are also triple-band phones in the GSM market that support 900, 1,800, and 1,900 MHz. Dual-band phones allow callers to access different frequencies in the same or different geographic regions, essentially giving their phone a wider coverage area.

Dual Mode   Dual-mode phones are those that support more than one technology. Typically, this is either CDMA and AMPS or TDMA and AMPS, but other dual-mode phones are starting to appear on the market, such as GSM and TDMA.

E911   A service mandated by the FCC for U.S. mobile carriers. The service will allow location information for subscribers calling 911 to be transmitted automatically to a public safety answering point (PSAP).

EDGE   Enhanced data rates for global evolution. A 3Glite standard transmitting data at a high rate (up to 384kbps), based on packet technology being developed for TDMA-based networks (IS-54, IS-136, and GSM).

Enhanced Data   See EDGE.
  Rates for Global
  Evolution

Enterprise Resource   See ERP.
  Planning

ERP   Enterprise resource planning. An acronym describing
                           corporate business systems that are fully integrated and
                           coordinated with each other. For example, a customer
                           order to a salesperson is linked to an order for factory
                           production, which is linked to an order for more parts. It
                           includes sophisticated reporting tools.

eXtensible Markup   See XML.
  Language

General Packet   See GPRS.
  Radio Service

Global System for   See GSM.
  Mobile
  Communications

GPRS   General packet radio service, also known as 2.5G,
                           deployed by AT&T, T-Mobile, and Cingular. A packet-
                           switched data technology that is being primarily deployed
                           for GSM networks.

GSM   Global system for mobile communications. A 2G standard
                           used throughout Europe as well as countries in the Middle
                           East, Asia, Africa, South America, Australia, and North
                           America. GSM's air interface is based on dividing frequency
                           bands into time slots (similar to TDMA). The upgrade for
                           GSM networks is GPRS.

Handheld Device   See HDML.
  Markup Language

HDML   Handheld device markup language. It uses hypertext
                           transfer protocol (HTTP, the underlying protocol for the
                           Web) to allow for the display of text versions of Web pages
                           on wireless devices. Unlike wireless markup language,
                           HDML is not based on XML. HDML also does not

allow developers to use scripts, whereas WML employs its own version of JavaScript. Phone.com, now part of Openwave Systems, developed HDML and offers it free of charge. Website developers using HDML must recode their Web pages in this language to tailor them for the smaller screens of handhelds.

HTML   Hypertext markup language. The most common language used to present information on an Internet browser. This scripting language describes data and where the data are placed on a page.

HypterText Markup   See HTML.
  Language

iDEN   A modified TDMA technology used by Motorola and run by Nextel Communications, Southern LINC, and a handful of other carriers around the world. iDEN phones run on a different frequency from other cellular services and are therefore incompatible with them.

IEEE   Institute of Electrical and Electronics Engineers. The IEEE is a nonprofit technical professional association that promotes electronic ideas and standards both in the United States and worldwide.

IM   Instant messaging. Messaging with presence (ability to view who is present). IM is capable via all network types.

I-MODE   A very popular service in Japan for transferring packet-based data to handheld devices. It is based on a compact version of HTML and does not use WAP standards. AT&T Wireless and the creator of I-Mode, NTT DoCoMo, may bring the I-Mode service to the United States in the future.

Instant Messaging   See IM.

Institute of   See IEEE.
  Electrical and
  Electronics
  Engineers

Internet Protocol   See IP.

IP   Internet protocol. An industry term for communications that flow through the Internet using routing and origination addresses.

Kbps   Kilobits per second.

LAN   Local area network. A computing network covering short distances, such as within a building.

Latency   A cool techie word for the measure of diminished running speed of an application over a network. There are lots of reasons why wireless devices have high latency, including transmission speed, file size, or processing power.

LBS   Location-based services. Services or applications that center around a user's location in a mobile environment. Location-based services utilize location-sensitive technology, such as global positioning system (GPS) or network-based solutions, to deliver services or applications to a wireless device such as a mobile phone. These services can include finder applications that let mobile phone users locate friends or family, businesses, or landmarks. They can also deliver maps, directions, or traffic reports.

Local Area Network   See LAN.

Location-Based Commerce   Location-based commerce (L-commerce) refers to commercial transactions that take place in a mobile environment but are dependent, in some way, on the physical location of the customer or the physical location of a business. This technology makes it possible to pinpoint a person's whereabouts by locating his or her vehicle, cell phone, or other wireless device. A rescue team can locate where a crashed vehicle swerved off the road or a business can know when you are in the area and push you the lunch special for the day. Lots of issues with privacy. Related: e-911.

Location-Based Services   See LBS.

M-Commerce   Mobile commerce or mobile electronic commerce refers to commercial transactions and payments conducted in an untethered, non–PC-based environment. Transactions are made using wireless devices that can access data networks and send and receive information, including personal financial information.

Micropayments   Small payments that are typically aggregated by an m-wallet provider or other payment processor. Cahner's In-Stat/MDR considers payments from $.01 to $2.00 to be micropayments.

Millions of Instructions per Second   See MIPS.

MIPS   Millions of instructions per second. This is the way you would measure and compare the processing speed of a mobile computing device. For example, the Palm V runs at a maximum of 8 MIPS, the Window CE devices max out at 100 MIPS, and the Compaq IPAQ cruises at 200 MIPS. So, if you want more complex applications like Excel, video, and voice recognition to run in a reasonable amount of time, you need processing power—lots of it. New chips are being introduced to leapfrog to 500 or more MIPS.

MMS   Multimedia messaging service. A type of messaging comprising a combination of text, sounds, images, and video.

Mobile Commerce (or Mobile Electronic Commerce)   See M-Commerce.

Mobile Virtual Network Operator   See MVNO.

Mobile Wallet   See M-Wallet.

Multimedia Messaging Service   See MMS.

MVNO   Mobile virtual network operator. A company that, to end users, appears to be a wireless network operator. Unlike a standard wireless carrier, however, an MVNO does not own the base stations subsystem (BSS) that mobile network

operators (MNOs) do. MVNOs also may not necessarily
own other infrastructure one normally associates with a
MNO. More important, MVNOs do not hold licenses to
radio spectrum; instead they purchase network capacity
from wireless carriers that do hold licenses and that do
operate the network infrastructure necessary for wireless
phone communication.

M-Wallet

M-wallets, or mobile wallets, are software applications
that hold a user's sensitive personal and financial information, such as credit card numbers, expiration dates, bank
account information, passwords, and personal identification numbers (PINs). Most m-wallets are server-based,
which theoretically is more secure and avoids placing
data onto mobile devices, which are often processor- and
memory-constrained.

Packet

A unit of transmission over a network. The data to be
transmitted are split into packets, which are sent individu-
ally over the network.

PCS

Personal communication services. A general category for
two-way digital networks with integrated voice, data, and
messaging capabilities.

PDA

Personal digital assistant. Mobile, handheld devices, such
as the Palm series and Handspring Visors, that give users
access to text-based information. Users can synchronize
their PDAs with a PC or network; some models support
wireless communication to retrieve and send e-mail and
get information from the Web.

**Personal
Communication
Services**

See PCS.

**Personal Digital
Assistant**

See PDA.

Protocol

A specification of the interaction between systems and the
data manipulated. This describes what to do and when
(the rules), and the format of the data exchanged on the
lower communication layer.

Radio Frequency Devices	These devices use radio frequencies to transmit data. One typical use: A barcode scanner gathers information about products in stock or ready for shipment in a warehouse or distribution center and sends it to a database or ERP system.
Radio Frequency Identification Device	See RFID.
RFID	Radio frequency identification device. Shipping containers, equipment, or even pallets are equipped with RFID tags that can be read with a handheld reader pointed at the tag. The RFID tags are as small as a lighter or as large as a brick and can hold as much as 128K in information, including shipping history and contents. The tags can also be fitted with GPS technology so they can be tracked from the sky.
Short Messaging Service	See SMS.
Smart Phone	A combination of a mobile phone and a PDA, smart phones allow users to converse as well as perform tasks, such as accessing the Internet wirelessly and storing contacts in databases. Smart phones have a PDA-like screen. As smart phone technology matures, some analysts expect these devices to prevail among wireless users. A PDA equipped with an Internet connection could be considered a smart phone. Ericsson, Nokia, and Motorola also make smart phones.
SMS	Short messaging service. The quasi–email feature that allows short text messages (maximum 15 characters) to be sent and received via mobile phones. Somewhat popular in Europe and questionable for the United States, it has limited business application because of the limitation of the message length (i.e., short!).
Text-to-Speech	See TTS.
TDMA	Time division multiple access. A technique used to share the same bandwidth among different channels using periodic time slots. TDMA divides frequency bands available

to the network into time slots, with each user having access to one time slot at regular intervals. TDMA thereby makes more efficient use of available bandwidth than the previous generation AMPS technology. Used on either 800 or 1,900 MHz frequency bands.

Time Division Multiple Access

See TDMA.

TTS

Text-to-speech. The flip side of speech recognition, TTS takes written words and converts them to speech. Thus, when a caller requests specific information from a voice portal, such as driving directions, TTS reads the directions to the caller. Early TTS efforts were slow and were usually read by a computerized voice that was often referred to as Igor because of its similarity to the voice of the character of the same name in old horror movies. Current TTS technology is much more natural sounding, and in some situations the caller would be challenged to differentiate TTS from an actual human speaker.

Voice eXtensible Markup Language

See VXML.

Voice over IP

See VoIP.

Voice Portal

A voice portal is a software application that uses speech recognition technology to provide information to callers. Using a combination of speech recognition and text-to-speech technology, the application lets callers request specific information, such as news, weather, traffic reports, or e-mail, which is read by the application to the caller. Voice portals can also allow callers to conduct transactions, such as trade stock or manage bank accounts. Callers can also use voice portals to purchase products or services. Voice portals essentially allow callers to perform functions that they might otherwise do using the Internet or other methods. Additionally, the application can be used to authenticate callers by matching their voiceprint to one on file, for security purposes.

VoIP

Voice over Internet protocol. The process of taking voice (analog) and digitizing it (packetizing), then routing the packet via the Internet to the intended recipient and converting the digital packet back to analog for a voice conversion. This is the future for long distance traffic to the home (desktop).

VXML   Voice extensible markup language. The standard Internet markup language for use in speech applications. It allows voice portal applications to access Internet content and read it to callers.

WAG   Wireless application gateway. A name for the software that acts as middleware for companies wanting to extend their enterprise systems to wireless. Fifty companies, including Oracle, Microsoft, and IBM, teamed up to write a standard for software packages, including prewritten interfaces to most mobile devices and APIs (preset hooks) into standard legacy systems and servers.

WAP   Wireless application protocol. A standard or protocol for wireless devices and the accompanying infrastructure equipment. WAP provides a standard way of linking the Internet to mobile phones, PDAs, and pagers/messaging units.

WCDMA   Wideband code division multiple access (true 3G speeds). Very different from CDMA or CDMA2000. It is based on the GSM standard (UMTS), which is the most popular worldwide by a large and growing margin. WCDMA is more expensive to deploy than CDMA2000 and in the United States may lose the battle for first place with 3G while it wins the war worldwide.

WEP   Wired equivalent privacy. An algorithm whereby a pseudo-random number generator is initialized by a shared secret key. When this encryption is incorporated into a wireless LAN, eavesdropping is made much more difficult.

**Wideband Code   See WCDMA.
Division Multiple
Access**

**Wired Equivalent See WEP.
Privacy**

**Wireless Application See WAG.
Gateway**

Wireless   See WAP.
Application
Protocol

Wireless LAN   WLAN. It uses radio frequency technology to transmit
                        network messages through the air for relatively short
                        distances, like across an office building or college campus.
                        A wireless LAN can serve as a replacement for or exten-
                        sion to a wired LAN.

Wireless Markup   See WML.
Language

Wireless Point-of- See WPOS.
Sale

Wireless Spectrum   A band of frequencies where wireless signals travel carry-
                        ing voice and data information. Wireless spectrum is
                        typically auctioned or assigned to carriers by each national
                        government.

WML   Wireless markup language. A version of HDML, WML
                        is based on XML and will run with its own version of
                        JavaScript. Wireless spectrum is typically auctioned or
                        assigned to carriers by each national government.

WPOS   Wireless point-of-sale. Wireless machine-to-machine
                        communication, not including traditional data or voice-
                        centric devices that allows for credit/debit card transactions.
                        Typically, receipt capabilities exist. This does not include
                        credit/debit card purchasing capabilities through a tradi-
                        tional voice-centric handset.

XML   Extensible markup language. A technology that is rapidly
                        becoming the global method of choice for creating Web
                        content. It operates over multiple devices and network
                        platforms.

Index